CW00548335

Flow Visualization

Second Edition

Flow Visualization

Second Edition

WOLFGANG MERZKIRCH

Universität Essen
Federal Republic of Germany

1987

ACADEMIC PRESS, INC.

Harcourt Brace Jovanovich, Publishers
Orlando San Diego New York Austin
Boston London Sydney Tokyo Toronto

ACADEMIC PRESS, INC.
Orlando, Florida 32887

United Kingdom Edition published by
ACADEMIC PRESS INC. (LONDON) LTD.
24–28 Oval Road, London NW1 7DX

Library of Congress Cataloging in Publication Data

Merzkirch, Wolfgang.
Flow visualization.

Bibliography: p.
Includes index.
1. Flow visualization. I. Title.
TA357.M47 1987 620.1'064 86-22279
ISBN 0–12–491351–2 (alk. paper)

PRINTED IN THE UNITED STATES OF AMERICA

87 88 89 90 9 8 7 6 5 4 3 2 1

Contents

Preface to the Second Edition

During the past decade there has been evidence of a rapidly growing interest in the methods of flow visualization. The tremendous progress in the development of new methods is reflected in the International Symposia on Flow Visualization of which the first took place in Tokyo in 1977. These symposia have contributed substantially to the interdisciplinary exchange between scientists and engineers using these methods, and they particularly make evident the great achievements made by our Japanese colleagues in this field. The second edition of this book tries to account for the new developments, applications, and results.

While preparing this book, I have benefited from many discussions with colleagues in the field and particularly from discussions with my students, who have been enthusiastically active in this area. Again, a great number of authors and research institutions have kindly provided photographs and illustrations, and I am very grateful for this help. I cordially thank Ms. Barbara Brandt, who typed the manuscript, as well as the publisher, who showed great patience when I was preparing this second edition.

Preface to the First Edition

Flow visualization is one of many available tools in experimental fluid mechanics. It differs from other experimental methods in that it renders certain properties of a flow field directly accessible to visual perception.

In this book an attempt is made to describe and review the most widely used methods for visualizing flows. To define the range of the subject I have chosen as a criterion the test as to whether a certain technique would enable one to obtain a visible flow pattern that could be investigated by visual inspection. This excluded, although optical, all spectroscopic methods. The range of topics still remains very broad, encompassing simple dye techniques, density-sensitive visualization methods, and application of electron beams and streaming birefringence. I have tried to distinguish the various methods and the range of their applicability by outlining the physical principles on which each is based.

For the purpose of getting more technical information, the attention of the reader is drawn to the references contained in books, journals, and other official publications. This list must of course remain incomplete. I shall be indebted to readers who draw my attention to other relevant references which I might have missed.

Another important source of information concerning the techniques of flow visualization are the scientific films produced and distributed by, e.g., the National Committee for Fluid Mechanics Films or by research institutions such as the National Aeronautics and Space Administration (NASA) and Office National d'Etudes et de Recherches Aérospatiales (ONERA).

An essential requirement for the presentation of this subject matter is to illustrate the various methods with appropriate flow pictures. I asked a great number of scientists and research institutions for such illustrations, and they all kindly agreed to contribute the requested material. I am therefore indebted to all these authors and institutions for having made available their records for reproduction in this book. I also wish to thank Professor R. J. Emrich and Dr. V. Vasanta Ram for many helpful and stimulating discussions. I am thankful to my wife and my family for their patience and support while I was studying the subject.

This book is dedicated to the memory of the late Hubert Schardin who fascinated his students with his continuing interest in making the invisible visible.

1

Introduction

1.1. Principles of Flow Visualization

The insight into a physical process is always improved if a pattern produced by or related to this process can be observed by visual inspection. This becomes obvious if we think of a fluid-mechanical process where a fluid is flowing in a channel or around a solid obstacle. By observing such a flow pattern, which might be stationary or variable with time, one can get an idea of the whole development of the flow. However, most fluids, gaseous or liquid, are transparent media, and their motion remains invisible to the human eye during a direct observation. In order to be able to recognize the motion of the fluid, one must therefore provide a certain technique by which the flow is made visible. Such methods are called flow-visualization techniques, and they have always played an important role in the understanding of fluid-mechanical problems. However, besides such instructive applications, the greater importance of many flow-visualization techniques is that one can derive quantitative data from the obtained flow picture. Such techniques provide information about the complete flow field under study without physically interfering with the fluid flow. In contrast, a single flow-measuring instrument, such as a certain pressure or temperature probe, provides data for only one point in the flow field, and in addition, the fluid flow is disturbed to a certain degree owing to the physical presence of the measuring probe. The experimental evaluation of some fluid-mechanical phenomena, for example, the origin and development of turbulence, still suffers from the fact, that the interference between the flow under study and the measuring device significantly affects the experimental results. In order to overcome these fundamental difficulties there is a continuous development to reduce the physical size of the measuring probes, and to improve, on the other hand, the quantitative evaluation characteristics of flow-visualization or optical techniques.

A great variety of flow-visualization experiments have been performed by Reynolds and Prandtl in order to support and illustrate their pioneering research in fluid mechanics. However, if one tries to find a person who has played a pioneering role for the development of flow visualization, one would think foremost of Ernst Mach who was already familiar with schlieren and interferometer techniques, high-speed photography, and many other methods. His approach to flow visualization was more fundamental than from a simple engineering standpoint only. Mach's interest in flow visualization must be seen in connection with his role as a phenomenologist, who believed that (visual) sensations are the source of scientific evidence. Flow-visualization experiments were a concrete support of Mach's phenomenological standpoint, which has been characterized by R. S. Cohen in the following way: "Mach wished science to assert only that for which there can be evidence. Sensed data are the common shared mode of evidence for every observers. Mach took it for granted that knowledge was sensational."[1] And Ernst Mach himself describes his views: "Modern science strives to construct its picture of the world not from speculations but so far as possible from facts. It verifies its constructs by recourse to observation."[2]

The methods of flow visualization can be classified roughly into three groups, which coincide with the subjects discussed in Chapters 2–4. The first class comprises all techniques by which a foreign material is added to the flowing fluid that might be gaseous or liquid. The foreign material must be visible, and if the particles of which the material is composed are small enough, one may assume that the motion of these particles is the same as that of the fluid, in direction and magnitude of velocity. The visualization is thus an indirect method, since one observes the motion of the foreign material instead of the fluid itself. The difference between the movement of the fluid and that of the foreign particles can be minimized, but not totally avoided, by giving to the particles a density almost coinciding with that of the fluid. These methods give excellent results in stationary flows,

[1] R. S. Cohen (1970). Ernst Mach: Physics, Perception and the Philosophy of Science. *In* "Ernst Mach, Physicist and Philosopher," Boston Studies in the Philosophy of Science, Vol. VI. Reidel, Dordrecht–Holland.

[2] E. Mach (1943). "Popular Scientific Lectures," transl. by T. J. McCormack, 5th ed. Open Court, La Salle, Illinois.

but the errors can be enormous for unsteady flows, owing to the finite size of the particles. The methods also fail to give precise results, if the thermodynamic state of the fluid varies in the flow field as in flows with variable density (e.g., compressible flow). The thermodynamic properties of the foreign material are usually quite different from those of the fluid, although their density might be of a comparable value, and a change of the thermodynamic state causes relaxation phenomena, particularly a difference in the mechanical motion of fluid and foreign material. The variation of the fluid density is at the same time the key to the second class of visualization methods. Since the fluid density is a function of the refractive index of the flowing medium, compressible flows can be made visible by means of certain optical methods that are sensitive to changes of the index of refraction in the field under investigation. The flow field with varying density is, in optical terms, a phase object; that is, a light beam transmitted through this object if affected with respect to its optical phase, but the intensity or amplitude of the light remains unchanged after the passage of the object. An optical device behind the object provides in a recording plane a nonuniform illumination, according to the phase changes caused by the object. From the pattern in the recording plane one can make conclusions concerning the density variations in the flow field.

The ranges of applicability of the two visualization principles described so far, the addition of foreign material and the optical methods, coincide approximately with the classes of incompressible and compressible flows, respectively. One can now distinguish a third group of visualization techniques that is somehow a combination of the two above-mentioned principles. In this case, the foreign substance introduced into the flowing fluid is energy (e.g., in the form of heat or electric discharge). The fluid elements thus marked by their increased energy level sometimes need an optical visualization method so that they can be discriminated from the rest of the fluid. In other cases the energy release is so high that the marked fluid elements become self-luminous and can be directly observed. These methods are often applied to flow with a low average density level. Density changes occurring in such flows can be too weak to be detected by an optical method. Hence, this third group of visualization techniques is at least partly applied to a third class of flows, which one often distinguishes from the "ordinary" incompressible and compressible flows, namely, the class of rarefied or low-density gas flows. One must be aware that this third visualization technique is not a nondisturbing method, since it affects, more or less, the original flow according to the amount of released energy.

The application of flow-visualization techniques covers a wide field in engineering sciences and experimental physics. The discussion of the techniques and methods in Chapters 2–4 includes some of such applications, and for several reasons it was felt that the main emphasis should be given to the physical background of each method. This should enable the reader to recognize and to decide what method might be suitable for study of a given flow problem. Further possibilities of applications of the individual visualization methods as well as greater technical details can be found in the cited references. The various survey articles and books on experimental fluid mechanics listed at the beginning of the bibliography are also helpful for this purpose.

1.2. Flow Visualization and Optical Testing Methods

Among the experimental methods for testing fluid flows one often distinguishes between optical and nonoptical ones. An optical method does not necessarily provide visualization of the flow, at least in the sense as we have introduced this term here. It is interesting to discuss the role of flow visualization in the framework of optical testing methods. These methods usually are referred to as being nonintrusive or nondisturbing. With this negative reference one expresses that, in applying optical test methods, one avoids the introduction of a mechanical probe into the flow; probe measurements are inevitably associated with the generation of disturbances in a flow field, even if the probe is made mechanically extremely small. This explains the growing interest in optical testing methods. Any optical method is based on the interaction of a light wave with a fluid flow; the light is modified due to this interaction and carries information on the state of the flow. In most cases, the respective modification of the fluid is not measurable, and this is why optical methods are considered as nondisturbing. Information on the interaction process can be obtained in two different ways (Fig. 1.1):

1. One may receive the light transmitted through the fluid and compare its state with that of the incident light. The obtainable information then is integrated along the whole path of the light in the fluid.
2. One may record the light that is scattered from a certain position in the fluid into a specified direction. A general assumption is that this light carries information on the state of the flow at the position where this light is scattered from, and that this scattered light is not changed during its further passage through the fluid. Hence, the recorded information is local, which is of importance if the fluid flow is three-dimensional. The intensity of the scattered light is always much lower than that of the transmitted light.

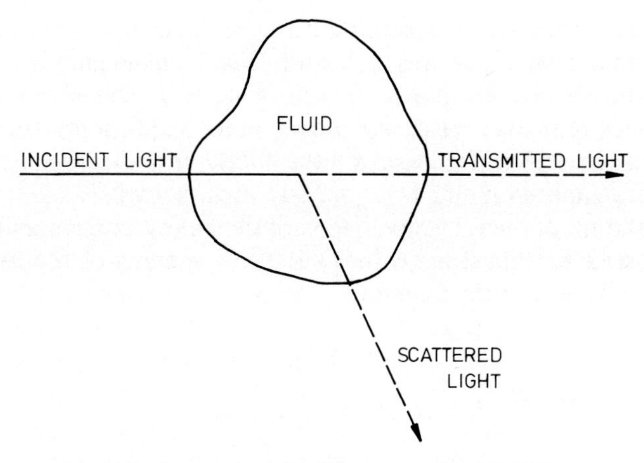

Fig. 1.1 Interaction of a light wave with a fluid flow.

The light transmitted through the flow field is recorded and analyzed, and the properties of this radiation can be compared with the respective properties of the incident (or undisturbed) radiation. The properties of a light wave are described by a number of quantities. Each of these quantities may be changed due to the interaction process. The measuring methods are sensitive to the changes of one of these quantities, and the following classification can be made:

Quantity changed and measured	Measuring method
Amplitude	Extinction measurement
Direction	Shadowgraph, schlieren
Frequency	Doppler schlieren
Phase	Interferometry
Polarization	

Extinction measurements are applied to two-phase flows where the presence of one phase (e.g., solid) obscures the transmitted light beam. The degree of extinction is described by Beer's law, which allows for concluding onto the concentration of the absorbing phase in the transparent phase.

Light is deflected from its original direction due to refraction in the fluid (i.e., due to changes of the refractive index). The deflection angles, which are made visible by means of the shadowgraph or the schlieren method are very small in gases, only a fraction of a degree, but they can be appreciably large in liquids.

It is known that the frequency of a light wave is shifted due to the Doppler effect. This also applies to the light deflected from a moving phase object and visualized in a schlieren system. A system that combines the principles of the laser Doppler anemometer and those of the schlieren system has been used to measure the velocity of flame fronts and shock waves in transmitted light.

The variation of the refractive index of the fluid also changes the phase of the transmitted light wave with respect to a wave, which would propagate in the absence of the fluid flow. The phase changes can be visualized and measured quantitatively by means of optical interferometers. Their principles and application as well as those of the former methods will be discussed in Chapter 3.

The direction of polarization of a linearly polarized light wave is changed if the wave propagates through a liquid sugar solution. This effect is applied to the measurement of sugar concentration in water; no use of this effect has been made for studying a flow field.

In all of the mentioned methods or applications the information on the state of flow is integrated along the path of the transmitted light. For the purpose of quantitative evaluation one has to desintegrate the recorded data, a problem that is known also in other fields of experimentation (e.g., spectroscopy or x-ray examination). If one wishes to resolve the distribution of the fluid parameters in a random, three-dimensional flow field, one has to direct the light in a number of different directions through this field. The set of recorded data is used in a numerical procedure, which by means of solving a set of integral equations, delivers the fluid parameters as functions of the three spatial coordinates (x, y, z). Procedures of this kind are known as tomography (see, e.g., Sweeney and Vest, 1973; Santoro *et al.*, 1981). The situation is less complicated if the flow field is axisymmetric. Then, only one direction for transmitting the light through the flow, normal to the axis of symmetry, is needed; and the respective integral equation has an analytic inversion (''Abel inversion'').

That portion of light which is not transmitted further through the fluid (i.e., which is ''extincted'') is reflected or scattered from the fluid particles. Either the fluid molecules or tracer particles, with which the fluid is seeded, can act as scatterers. The radiation scattered from (liquid or solid) tracers is much more intense than that from molecules, because the intensity of this radiation depends strongly on the size of the scatterer. Each scatterer has its specific characteristics: Both intensity and polarization of the scattered light depend on the direction into which this light is emitted. The frequency of this radiation can change as a function of the state of motion (Doppler effect) or the thermodynamic state (Raman effect) of the scatterer. Finally, one can distinguish between elastic scattering without any energy exchange between the incident radiation and the scatterers, and inelastic scattering where such an energy exchange takes place, thereby changing the internal energetic state of the scatterers.

In the latter case, the frequency of the emitted radiation usually differs from that of the incident radiation. The dependence of the various optical methods on the different scattering processes is listed in the following scheme:

Scattering	Elastic	Inelastic
From molecules	Rayleigh scattering	Raman scattering
From tracer particles	Doppler velocimetry; laser-dual-focus; speckle photography; particle tracking; flow visualization	Fluorescent tracers

The intensity of the elastic Rayleigh scattering is proportional to the number density of the scatterers in the measuring volume (Escoda and Long, 1983). Fluid density measurements can be made as long as the fluid constituency is known. Temperature measurements are possible if the Gaussian profile of the scattered light can be observed. Rayleigh scattering cannot be used to identify various components in a fluid mixture, but it is the most intensive of the molecular light-scattering techniques, much stronger than the inelastic Raman scattering, which indeed allows for discrimination between the radiation originating from the molecules of the different species in a mixture. Since the Raman signal also gives information on the gas temperature, this inelastic scattering technique is most appropriate for studying flames and combustion processes (Lederman *et al.*, 1979; Lapp and Penny, 1977).

Flow visualization, by means of observing the light scattered from smoke or dye, is mainly a qualitative method. If one resolves the light scattered from single particles, the observation becomes quantitative and allows for measuring the velocity of the scatterer, which in many cases, can be taken as the fluid velocity. The listed methods differ by the techniques of recording the scattered radiation: laser Doppler velocimeter and laser-dual-focus method measure the velocity in one point of the flow (“optical probe”), while photographic particle tracking and speckle photography deliver data of the whole field.

The signal-to-noise ratio in such quantitative methods could be improved if the tracer particle would not just scatter but emit its own, specific radiation. This principle is realized by means of fluorescent tracers, usually iodine, with which a flowing gas is seeded. Bright fluorescence of this tracer material is induced if one illuminates the fluid with a characteristic radiation (e.g., with that of an argon laser).

One can distinguish between two types of *recording systems,* which provide different kind of information on the fluid flow. The *whole field* can be observed with the eye, with a photo- or movie camera, or with a television camera. The information then is continuous in space, but discontinuous in time, because records of the whole field are taken at specific instants of time, t_i. The information is recorded in the plane of observation, say the x-y plane, while z is the direction along which the optical signal propagates toward the recording plane. In the transmitted-light system the recorded data $\mathbf{J}$ is integrated along the path z:

$$\mathbf{J} = J[x(z), y(z), t_i]\, dz,$$

and, as stated earlier, the wanted information $J(x, y, z, t)$ only becomes available by solving respective integral equations. If one records the data scattered from a plane z_j = const in the flow, the recorded information can be expressed by

$$J = J(x, y, z_j, t_i).$$

The dependence on z can be found by taking successive records with varying values of z_j.

If the optical signal is received by an electro-optical sensor (e.g., a photomultiplier), the information is available as a continuous function of time, but only for one position (x_i, y_i, z_i)

$$J = J(x_i, y_i, z_i, t).$$

This way of recording optical data for one position is called an *optical probe*. A whole field can be surveyed by scanning the field or by using an array of electro-optical sensors at a large number of positions (x_i, y_i) in the recording plane.

The usual issue of an experiment is to compare the recorded results with other experimental values or with the predictions from a theory, in many cases the results of a numerical experiment. It is a conventional form that such a comparison is based on a diagram of the flow parameter of interest, for example, the pressure at a specific position as function of time (Fig. 1.2). Recently developed techniques of image processing allow for comparing the results on a different level: Digital flow images can be computed from either optical, numerical or probe data. Similarity in the flow pattern computed from the data of the different sources often is taken as proof for good agreement of the results. Such an interpretation of the

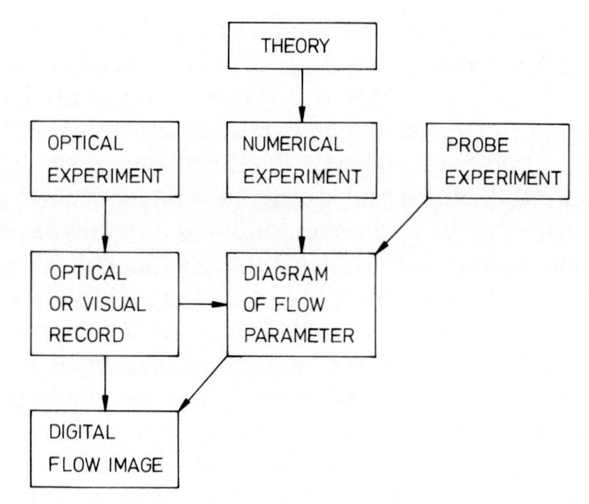

Fig. 1.2 Scheme and hierarchy of experimental methods.

results of image processing can be dangerous and misleading, mainly for two reasons:

1. A computed flow picture usually is the product of an integration process, and the result is not a unique solution; that is, different flow situations may generate the same (or almost the same) flow pattern.
2. Differences in the computed pattern are not linearly related to differences in the respective flow situations; that is, subjective similarity of two computed flow patterns might be associated with large objective differences in the related flows.

1.3. Image Processing

Image processing is the extraction and evaluation of data from a visible pattern by means of a computer. It requires the existence of a digitized form of the pattern. Though not being a new method of flow visualization, image processing has opened a new dimension for both qualitative and quantitative visualization experiments. Digitized flow pictures can be generated in different ways. As indicated previously, the results of computational flow studies can be displayed in form of computer generated graphics, which resemble physically taken flow photographs. A display in color improves the visibility or increases the amount of discernible information; as examples see Zabusky (1984) or Hung and Buning (1985). Processing of these graphics is not necessary because the data is already available as the result of the computation.

If one surveys a whole field with a probe, one may produce a map of the distribution of the respective flow quantity measured with the probe. The output of a probe measurement is in many cases an electric signal. This signal can be digitized and displayed on a screen where different levels of gray or different colors are attributed to particular levels of the respective signal. Crowder (1980) demonstrated that such planar flow field surveys and the associated display of a visible pattern can be realized with a total pressure probe moving in a plane normal to the mean flow. Related experiments in which isopressure maps were generated have been reported by Ostowari (1984), Crowder and Beck (1985) and Cogotti (1985), see Fig. 1.3, while Winkelmann and Tsao (1982) and Sakata *et al.* (1985) produced similar patterns based on the survey with a hot-wire probe. In all these cases, the data resulting in the visible pattern is not collected at the same instant of time, but during a certain period when the probe is traversing the flow field.

A real or physical flow visualization experiment may directly result in a digital image when an appropriate recording system is applied. This is, in the simplest case, a television camera whose output can be digitized and stored in the memory of a computer. More refined systems, particularly for fast recording and low signal intensities, have been reviewed by Winter *et al.* (1987); this review includes vidicon-based imaging systems and photodiode arrays.

Once the flow picture is digitized, it can be subjected to a computer analysis in which specific structures in the pattern are identified, provided that such structures exist. Typical examples are the fringes in an interfer-

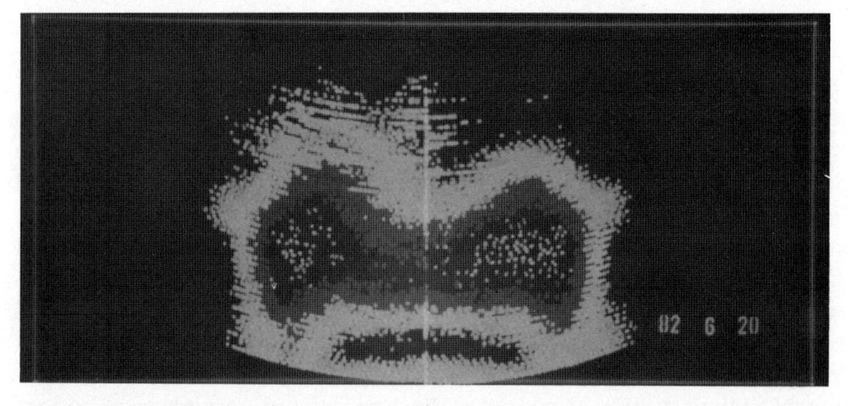

Fig. 1.3 Visualization of isopressure regions in the wake behind a car model tested in a wind tunnel. The field is mapped with a traversing total pressure probe. (From Cogotti, 1985. Published by Hemisphere Publishing Corporation.)

ogram, the time lines in hydrogen-bubble visualization, or simply the images of tracer particles. The recognition of the structures usually is based on a spectral analysis of the digital pattern. The coordinates of the structures or of their edges are determined, and the structures may be approximated by geometrical curves. Since the visible pattern is associated to the distribution of a certain flow quantity in the plane of view, this distribution can now be determined. This may be the fluid density in the case of interference fringes, in the two other examples it is the flow velocity. In Section 3.3.4. we give a number of references for computer-based evaluation of optical interferograms. Image processing of hydrogen-bubble time lines has been described by Lu and Smith (1985), and the computerized evaluation of tracer particle photographs with the subsequent display of isovelocity contours is illustrated, e.g., by Utami and Ueno (1984), Kobayashi *et al.* (1985), or Balint *et al.* (1985). Utami and Ueno even show how to compute a flow pattern that would have been physically generated with another flow visualization technique that, in reality, has not been applied; they compute a pattern, from the tracer particle-data that would have been produced in the same flow by means of the hydrogen-bubble technique.

Interference fringes and time lines are distinct structures, purposely produced by particular visualization methods. The flow itself might develop internal structures that need to be visualized (e.g., shock waves, wakes, or vortices). They can easily be identified in a laminar flow. In turbulent flow, such large-scale vortices, often designated as coherent structures, are masked by the small-scale turbulent noise. Corke (1984) shows how to search for coherent structures in a digitized flow picture by applying match filtering. The aim is the reconstruction of low wave number features, which is accomplished by filtering the image with a two-dimensional low-pass filter function (Fig. 1.4). In this way, the recognition

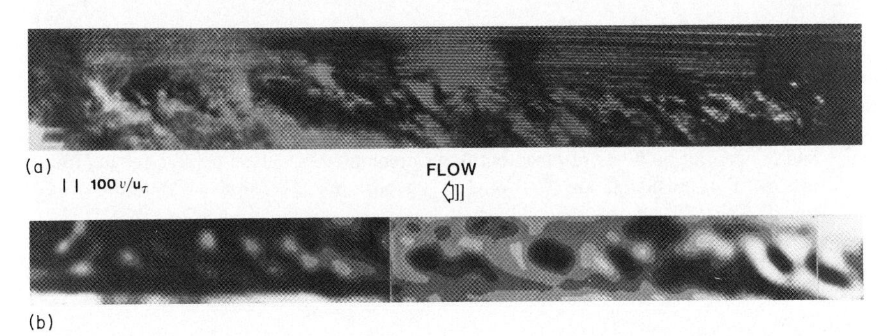

Fig. 1.4 Digital realization of flow visualized in a turbulent boundary layer (a) and corresponding matched filtered representation (b) with suppression of large-scale intermittancy. (From Corke, 1984. Copyright © American Institute of Aeronautics and Astronautics; reprinted with permission.)

of a particular pattern is enhanced, and it is made independent of a subjective interpretation. The latter must be regarded to be one of the great achievements of image processing, besides the possibility of an automated evaluation of the recorded visible pattern.

Bibliography

Books, Reviews, and Surveys

Asanuma, T., ed. (1977). "Handbook of Flow Visualization." Asakura Shoten, Tokyo. (In Jpn.)

Asanuma, T., ed. (1979). "Flow Visualization," Proceedings of the First International Symposium on Flow Visualization, Tokyo. Hemisphere, Washington, D.C.

Flow Visualization Society of Japan (1984). "Collection of Flow Visualization Photographs," No. 1. Tokyo. (In Jpn.)

Flow Visualization Society of Japan (1985). "Collection of Flow Visualization Photographs," No. 2. Tokyo. (In Jpn.)

Journal of the Flow Visualization Society of Japan (1981). Starting with Vol. 1. (In Jpn.)

Merzkirch, W. (1979). Making fluid flows visible. *Am. Sci.* **67,** 330–336.

Merzkirch, W., ed. (1982). "Flow Visualization II," Proceedings of the Second International Symposium on Flow Visualization, Bochum. Hemisphere, Washington, D.C.

Nakayama, Y., ed. (1984). "Photo-Album of Flow." Maruzen, Tokyo. (In Jpn.)

Reznicek, R. (1972). "Visualisace Proudeni" (Flow Visualization). CSSR Acad. Sci., Prague. (In Czech.)

Settles, G. S. (1986). Modern developments in flow visualization. *AIAA J.* **24,** 1313–1323.

Van Dyke, M. (1982). "An Album of Fluid Motion." Parabolic Press, Stanford, California.

Yang, W. J. (1985). "Flow Visualization III," Proceedings of the Third International Symposium on Flow Visualization, Ann Arbor. Hemisphere, Washington, D.C.

References

Balint, J. L., Ayrault, M., and Schon, J. P. (1985). Quantitative investigation of the velocity and concentration fields of turbulent flows combining visualization and image processing. *In* "Flow Visualization III" (W.-J. Yang, ed.), pp. 254–258. Hemisphere, Washington, D.C.

Cogotti, A. (1985). A passenger car wake survey using coloured isopressure maps. *In* "Flow Visualization III" (W.-J. Yang, ed.), pp. 668–675. Hemisphere, Washington, D.C.

Corke, T. C. (1984). Digital image filtering in visualized boundary layers. *AIAA J.* **22,** 1124–1131.

Crowder, J. P. (1980). Quick and easy flow-field surveys. *Aeronaut. Astronaut.* **10,** 38–45.

Crowder, J. P., and Beck, H. M. (1985). Electronic wake imaging system. *In* "Flow Visualization III" (W.-J. Yang, ed.), pp. 208–212. Hemisphere, Washington, D.C.

Escoda, M. C., and Long, M. B. (1983). Rayleigh scattering measurements of the gas concentration field in turbulent jets. *AIAA J.* **21,** 81–84.

Hung, C.-M., and Buning, P. G. (1985). Simulation of blunt-fin-induced shock-wave and turbulent boundary-layer interaction. *J. Fluid Mech.* **154,** 163–185.

Kobayashi, T., Ishihara, T., and Sasaki, N. (1985). Automatic analysis of photographs of trace particles by microcomputer system. *In* "Flow Visualization III" (W.-J. Yang, ed.), pp. 231–235. Hemisphere, Washington, D.C.

Lapp, M., and Penny, C. M. (1977). Raman measurements on flames. *Adv. Infrared Raman Spectrosc.* **3,** 204–261.

Lederman, S., Celentano, A., and Glaser, J. (1979). Flow field diagnostics. *AIAA J.* **17,** 1106–1110.

Lu, L. J., and Smith, C. R. (1985). Image processing of hydrogen bubble flow visualization for determination of turbulence statistics and bursting characteristics. *Exp. Fluids* **3,** 349–356.

Ostowari, C. (1984). A rapid technique for measuring and visualizing the extent of separated flow. *Exp. Fluids* **2,** 67–72.

Sakata, K., Shindoh, S., and Yanagi, R. (1985). Experimental flow analysis with computer graphics on film-cooling flow field. *In* "Flow Visualization III" (W.-J. Yang, ed.), pp. 203–207. Hemisphere, Washington, D.C.

Santoro, R. J., Semerjian, H. G., Emmerman, P. J., and Goulard, R. (1981). Optical tomography for flow field diagnostics. *Int. J. Heat Mass Transfer* **24,** 1139–1150.

Sweeney, D. W., and Vest, C. M. (1973). Reconstruction of three-dimensional refractive index fields from multidirectional interferometric data. *Appl. Opt.* **12,** 2649–2664.

Utami, T., and Ueno, T. (1984). Visualization and picture processing of turbulent flow. *Exp. Fluids* **2,** 25–32.

Winkelmann, A. E., and Tsao, C. P. (1982). A color video display technique for flow field surveys. *AIAA Pap.* **82-0611-CP.**

Winter, M., Lam, J. K., and Long, M. B. (1987). Techniques for high-speed digital imaging of gas concentrations in turbulent flows. *Exp. Fluids* **5,** 177–183.

Zabusky, N. J. (1984). Computational synergéstics. *Phys. Today* **37**(7), 36–46.

2

Addition of Foreign Materials into Gaseous and Liquid Flows

The flow velocity is a vector, and in measuring it one must provide for determining both the magnitude and the direction of the velocity as a function of spatial position and time. In the following sections we describe methods for visualizing and measuring the velocity field by visual inspection or recording of flow images. These methods all have in common that portions of the fluid are marked or seeded with a foreign material, and that one draws conclusions on the velocity field from the motion of this observable foreign material. It depends on the applied technique whether one measures the velocity direction, its magnitude, or both. These principal differences, which apply in equal sense to liquid and gaseous fluid flows, will first be outlined before technical details of the various methods are discussed.

2.1. Visualization of Flow Direction and Flow Contours

2.1.1. Streamlines, Filament Lines, and Particle Paths

At discrete points of the flow field, dye, smoke, or single particles are released into the fluid in order to visualize its flow. This foreign material is swept along with the mean flow. The motion or the trajectory of this material can be measured by means of an appropriately illuminated photograph. In order to interpret the observed motion of the added material one has to distinguish between the following definitions.

Streamlines are the curves tangential to the instantaneous direction of the flow velocity in all points of the flow field. If we provide a two-dimensional drawing of a field of velocity vectors for a particular moment, these vectors, represented by small arrows, will be tangential to the respective streamlines. In a three-dimensional flow field, the streamlines are given by the relation

$$u:v:w = dx:dy:dz,$$

where u, v, w are the three components of the velocity vector in the coordinate directions x, y, z. No fluid is flowing across a streamline at the instant considered. Streamlines can be visualized by seeding the fluid with small particles and photographing the flow field with an appropriate and known exposure time, so that each particle appears as a streak in the picture. Thus, one obtains the magnitude and direction of velocity in selected points of the flow field. The streamlines then can be found by drawing the curves tangential to the particle streaks.

A *filament line* is the instantaneous locus of all fluid particles, which have passed through a particular, fixed point of the flow field. Thus, filament lines, which sometimes are also named streaklines, can be visualized by continuously injecting dye, or smoke, or another appropriate material into the flow from selected positions.

A *particle path* or particle trajectory, is the curve that a fluid particle traverses in the flow field as a function of time. The particle path contains the integrated time history of the motion of one single fluid particle. It can be visualized if one takes a long-time exposure record of the motion of one foreign particle, which has been introduced into the flow.

These three curves coincide if the flow field is stationary. But in a flow, which explicitly depends on time, the three types of curves are different from one another. Whether one visualizes streamlines, filament lines, or particle paths, with the aid of the foreign material added to an unsteady flow, depends on the choice of the points where the particles are introduced, on the rate of the particles released to the flow, and on the length of the exposure time if a record of the whole field is taken.

The differentiation between streamlines, filament lines, and particle paths can be illustrated by a simple analogy. Let us consider the flow of automobiles on a system of highways, and suppose that we take a photograph at night when the cars are illuminated. If there is only one illuminated car on the street we obtain with an open-shutter photograph (time exposure) a well-defined "particle path" of this particular car, which can be discriminated from the nonilluminated cars. If all cars are illuminated and if we take a photograph with an exposure time of finite length, we receive a representation of the velocity of all cars for that particular moment; from this exposure one may deduce the "streamlines". If the highway is crowded and if one takes a photo from an airplane at high altitude, so that one cannot distinguish between the individual cars, one may obtain a lightened curve, which is equivalent to a dye or filament line. Hence, the kind of curve obtained in the picture depends on the choice of the exposure time and on the number of illuminated or marked automobiles moving in this system.

There is one further point, which has to be considered in interpreting the recorded lines or curves: this is the reference system from which the flow is observed or photographed. The preceding description refers to a case where the observer is at rest in the laboratory system; that is, the observer is at rest with respect to the device from which the foreign material (dye or smoke) is released into the flow. It is also possible that the observer or the camera are in motion relative to such system. The recorded streamlines or filament lines then refer to a moving reference system, and so do the measured velocities. If the automobiles in the aforementioned example are photographed from a plane flying at constant speed in the mean direction of the flow, one may well discriminate between those cars that go faster or slower than this reference speed. Reversed flow may occur in such a system, and it is obvious that the shape of the streamlines or filament lines is different for each reference system from which the flow is observed. This explains that one must be cautious and thoughtful in interpreting such flow photographs. Hama (1962) has discussed in detail the shape of streamlines and filament lines in a sinusoidally perturbed shear flow, when the lines are recorded with a moving camera or with a camera at rest in the laboratory system (Fig. 2.1). Hama shows that this unsteady flow may appear steady to an observer who moves at the wave velocity with the flow. The apparent steadiness of this flow as seen from a moving reference system and the occurrence of reversed flow regimes in such system appears to be a typical phenomenon for periodically disturbed flows. This phenomenon has been observed also by Hung and Brown (1976) in the case of model studies of peristaltic flow.

If the velocity distribution in a whole field has been determined by means of one particular experimental method, it is possible to compute flow pictures of the same field, but under different recording conditions, for example, with a superimposed constant velocity, curves of constant swirl, or even a pattern like it would have been obtained with a different visualization technique (see, e.g., Utami and Ueno, 1984).

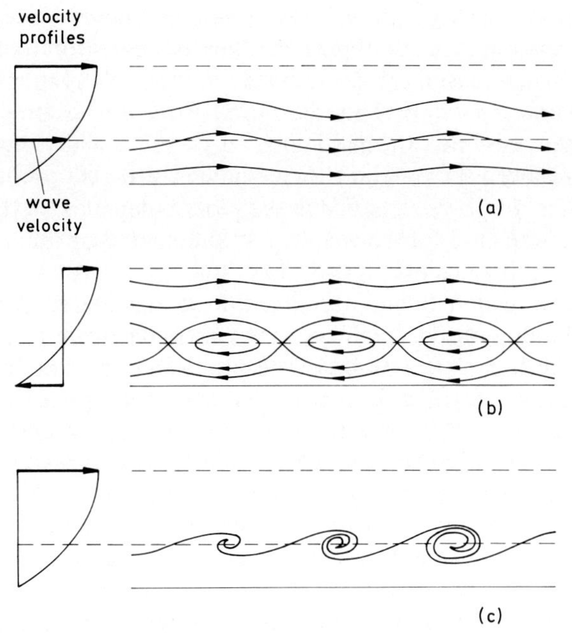

Fig. 2.1 Shear layer flow with traveling instability waves: (a) streamlines seen by stationary observer, (b) streamlines seen by observer traveling at the wave velocity, and (c) filament lines. (From Hama, 1962.)

2.1.2. *Dye Lines and Contours in Liquid Fluid Flows*

The marking of lines or contours in a flowing liquid by means of dye can be achieved by introducing the dye into the liquid from outside (''direct injection''), or by generating it with an appropriate chemical reaction in the liquid. In the second case it is required that the liquid carries respective chemicals in solution, and that the dye-producing reaction is initiated at the proper location in the flow. A number of such chemical techniques will be described later, but first, we will discuss the direct injection of dye and some general problems which are common to any dye visualization in a liquid flow.

The injection of dye has since long been a popular method for visualizing water flows. The dye is released either from a small ejector tube placed at a desired position in the flow field or from small orifices, which are provided in the wall of a certain model under investigation. In both cases one cannot avoid the fact that the main flow is, to a certain degree, disturbed by the presence of the ejecting device. The tube from which the dye is dispersed must be placed far enough upstream of a test model, so that interference of the tube with the flow pattern to be studied will be minimized. The released dye is contained in the wake of the tube, and one should prevent the wake of the dye ejector from becoming unstable or turbulent; this confines the technique to flows in which the tube wake Reynolds number is below the critical value. Tubes that normally serve as Pitot probes in air flows have been used as dye injectors (Offen and Kline, 1974). Such tubes are fabricated with an outer diameter of 1 mm or less; other injecting devices are hypodermic tubes or syringes.

The rate at which the dye is released has to be matched with the velocity of the liquid. If the injection rate from the injector tube is allowed to become too large, the issuing dye might behave like a jet, and vortices appear along the interfaces between the jet and the main flow. If the dye is released from small holes in the surface of a rigid test model, the velocity component of the dyed solution perpendicular to the model surface must be minimized; otherwise the flow of the injected dye would interfere with the main flow around the model, and particularly would alter the body wall boundary layer due to the injected mass and momentum.

A dye suitable for the visualization of filament lines or flow contours has to fulfill a number of requirements. Besides some general properties that apply to any foreign material used for flow visualization (e.g., non-toxic, noncorrosive) there are mainly three conditions the dye should meet, that is neutral buoyancy, high stability against mixing, and good visibility.

A dye is neutrally buoyant if it has the same specific weight as the working fluid, in most cases water. This value of the specific weight can be met by mixing the prepared dye with alcohol. Such a mixture is not a true solution, and, under the action of inertial or centrifugal forces, which act in different ways on the different components of the mixture, the dyed filaments will either decay or not indicate the true flow direction. The latter problem is minimized if the dyed solution can be prepared by complete dilution of the dye (e.g., food coloring or ink) in the working fluid (water).

As the dye propagates along a line in the flow it will mix with the surrounding fluid, and the dye lines will lose their clarity and rapidly decay, particularly in a turbulent flow. The mixing also occurs at the interface of differently dyed flow regimes or contours. Therefore, the method is restricted mainly to laminar flows or low fluid velocities. The mixing or diffusion of dye in the working liquid has not been investigated systematically. The dye filaments may be stabilized by mixing the dye with milk (Werlé, 1960; also demonstrated in many other publications of Werlé, e.g., 1973, 1976 (with Gallon), 1980), and it is presumed that the fattiness of the milk retards diffusion of the dyed solution into the main bulk of water. At the same time, due to its high reflective properties, milk also meets the condition of good visibility, so that the use of milk in a dye mixture combines two advantages: that of high contrast of the dye lines and that of stabilized filaments with respect to a rapid diffusion.

The dyed portion of the fluid should be illuminated from behind with a diffuse source [e.g., spot lights (Offen and Kline, 1974) or banks of fluorescent tubes (Hunt and Snyder, 1980)]. Less appropriate is an illumination from the side (e.g., normal to the direction of observation). In such a lighting system, which is used for observing the light scattered from single tracer particles, shadows might occur in the flow field, either from the lightened dye lines or from the test model under study.

The choice of dye depends on the particular conditions of a flow experiment, but apparently also on such arbitrary circumstances as rapid availability or cost. Food coloring mixed with milk appears to be the most popular dye in pure and salt water. This mixture can be diluted with the water from the operating channel to minimize density gradients, and filaments of different colors can be generated for promoting the identification of the spatial flow structure, e.g. in a swirling flow (Figs. 2.2 and 2.3). Table 2.1 lists a number of dyes, which have been used for flow experiments, together with the working fluid and the respective references.

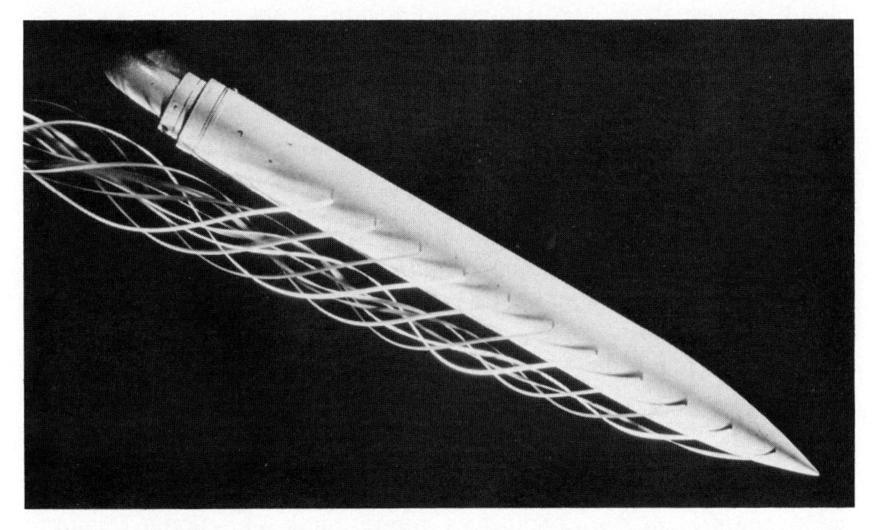

Fig. 2.2 Dye lines in the vortex flow behind a yawed cylinder. The fluid is water and the dye is a mixture of ink, milk, and alcohol. The color of the original dye lines is red, yellow, and blue. (From Fiechter, 1969.)

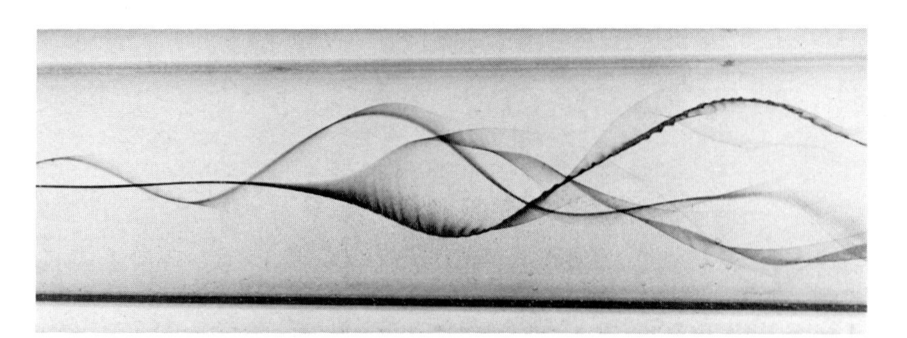

Fig. 2.3 Dye lines in swirling flow in a cylindrical tube. The fluid is water and the dye is food coloring. (From Sarpkaya, 1971.)

Further references can be found in the review articles by Clayton and Massey (1967) and Werlé (1973).

Utilizing a fluorescent dye is another means for improving the visibility of dye filaments, but this material also is well suited to mark larger fluid packets or flow regimes for a better differentiation from the rest of the fluid, even under such large-scale conditions as in an open lake (Csanady, 1963). These dyes must be excited to fluorescence by an appropriate light source, which radiates in the spectral range of greatest fluorescence of the dye (e.g., a mercury lamp or an argon laser). In this case, the illumination should be normal or oblique to the direction of observation, since one records the scattered fluorescent radiation. It is also obvious that the sensibility of the photographic film must fit the respective spectral range. The use of a fluorescing dye can allow for concentration measurements by means of a fluorometer (Lynch and Brown, 1974). The special role of fluorescent material for determining velocities or even the thermodynamic state in a flow will be discussed in a later section. Rhodamine, a nontoxic organic pigment is the most frequently applied dye of this type; 2% by weight are soluble in water, and it is also highly soluble in methanol, which might be of interest for adjusting the density. A number of different dyes and trade marks are known (see Table 2.1); Rhodamine–B emits a dark-red radiation when excited by an argon laser; Rhodamine–6G radiates yellow, Fluoresceine green.

The ejection of dye is a standard technique for experimenting with water tunnels. The dye ejected into a recirculating tunnel remains in the system and gradually contaminates the water. Dumas *et al.* (1981), who employed different fluorescent dyes, report that it takes about 25 min of continuous dye injection until the 5 m^3 water of the recirculating system have to be replaced. Many experimenters investigate and visualize aero-

TABLE 2.1

Dye Solutions Used for the Marking of Filament Lines and Flow Contours

Working fluid	Dye	References
Water	Milk and food coloring	Werlé (1960 f.f), Werlé and Gallon (1976), Deardorff *et al.* (1980)
Salt water	Milk	Simpson (1972)
Water	Food coloring	Han and Patel (1979), Maxworthy (1972), Offen and Kline (1974), Pullin and Perry (1980), Sarpkaya (1971)
Salt water	Food coloring	Hunt and Snyder (1980)
Polymer solution	Food coloring	Donahue *et al.* (1972)
Water	Ink	Faler and Leibovich (1977), Kang and Chang (1982)
Polyethylene/glycol solution	Ink	Masliyah (1972)
Water	Printer's white	Martin and Lockwood (1964)
Water	Potassium permanganate	Kotas (1977)
Water	Crystal violet	Hide and Titman (1967)
Water	Gentian violet	Sullivan (1971)
Water	Methylene blue dye	McNaughton and Sinclair (1966)
Salt water	Methynol blue dye	Withjack and Chen (1974)
Salt water	Blue Dextran dye	Delisi and Orlanski (1975)
Water	Rheoscopic fluid AQ 1000	Davis and Choi (1977)
Water	Black Nigrosin	Yeheskel and Kehat (1971)
Water	Rhodamine	Csanady (1963), Dumas *et al.* (1982), Gad-el-Hak *et al.* (1981), Lynch and Brown (1974)
Water	Fluoresceine	Dumas *et al.* (1982), Gad-el-Hak *et al.* (1981), Gaster (1969), Thomas and Cornelius (1982)
Water	Fluorescent ink (recorder ink)	Masliyah (1980)

dynamic flow phenomena in water tunnels (see, e.g., Werlé, 1973, 1979, 1982; Erickson, 1979). For the same unit Reynolds number and model scale, the velocity in water is $\frac{1}{15}$ of that in air, so that these phenomena can be observed at a relatively low speed. A special visualization technique has been developed for towing tanks. Thin, horizontal dye layers are produced in the water before the model is set into motion. The flow around the model displaces the colored layers whose motion can be observed and recorded. The dye layers can be stabilized by a weak saline stratification in the tank (Gad-el-Hak *et al.*, 1981), and, consequently, the technique is well suited for studies in stratified fluids (Delisi and Orlanski, 1975).

If one could instantaneously release the dye along a line perpendicular to the mean flow direction, it would enable one to visualize the local velocity profile by means of recording the distortion of this dye line in the flow. A mechanism for such a quick and simultaneous dye injection along all points of this line would represent a considerable disturbance in the flow, making this type of velocity measurement worthless. However, methods have been found to produce a dye in certain liquids along a prescribed curve and with exact temporal control, either by electrolytic or photochemical means. These dye techniques will be discussed separately (Section 2.3).

The generation and introduction of tracers and dyes, without mechanically disturbing the flow, has been the aim of many investigators. One way of nonintrusively marking the contour or boundary between two different flows is to generate a dye-producing, chemical reaction at the interface of the two flows. Most suitable for such a procedure is the use of pH-indicators, which can be solved in water and which change color upon a specific change of the pH-value. Quraishi and Fahidi (1982) describe a number of substances, which can be applied for this purpose (see Table 2.2). The pH-indicator could be dissolved into one of two streams, which subsequently are being mixed, e.g., in the fluid of a jet, which is exhausted into another fluid originally at rest. The two fluids are given different pH-values, below and above the critical value where the color

TABLE 2.2

pH Indicators and Range of Color Change[a]

Indicator	pH(1)/color	pH(2)/color
Bromo cresol green	3.6/yellow	5.2/blue
Bromo phenol red	5.2/yellow	7.0/red
Bromo thymol blue	6.0/yellow	7.6/blue
Meta cresol purple	7.6/yellow	9.2/purple
Thymol blue	8.0/yellow	9.6/blue
Phenolphthalein	8.3/clear	10/red

[a] From Quraishi and Fahidi (1982).

change takes place. Then, the mixing zone of the two fluids is visualized by the dye-producing reaction. Breidenthal (1979, 1981, 1982) has demonstrated that such mixing layers or shear layers can be visualized not only for qualitative means (Fig. 2.4), for example, for studying coherent structures, but that the mixing rate can be measured also quantitatively by applying a densitometer.

Another dye-producing reaction with time control has been developed for measuring mean flow times of a fluid through vessels of arbitrary shape (Denbigh *et al.*, 1962). The working fluid is prepared by mixing two chemical solutions. The time reaction consists of turning the mixture blue quite suddenly at a definite time after the instant of mixing. If the mixture is made up continuously and immediately before it enters a certain flow system, one may well recognize those parts of the fluid, which at any instant have been in the system for more than the length of time specified by the time reaction. Danckwerts and Wilson (1963) have used such a reaction to study the flow through cylindrical vessels with differently arranged inlets and outlets. By choosing the reaction time equal to or larger than the mean residence time of the fluid in the vessel, they could visualize regimes of reversed flow. The mixture applied by Danckwerts and Wilson contained sodium persulfate ($Na_2S_4O_8$), potassium iodide

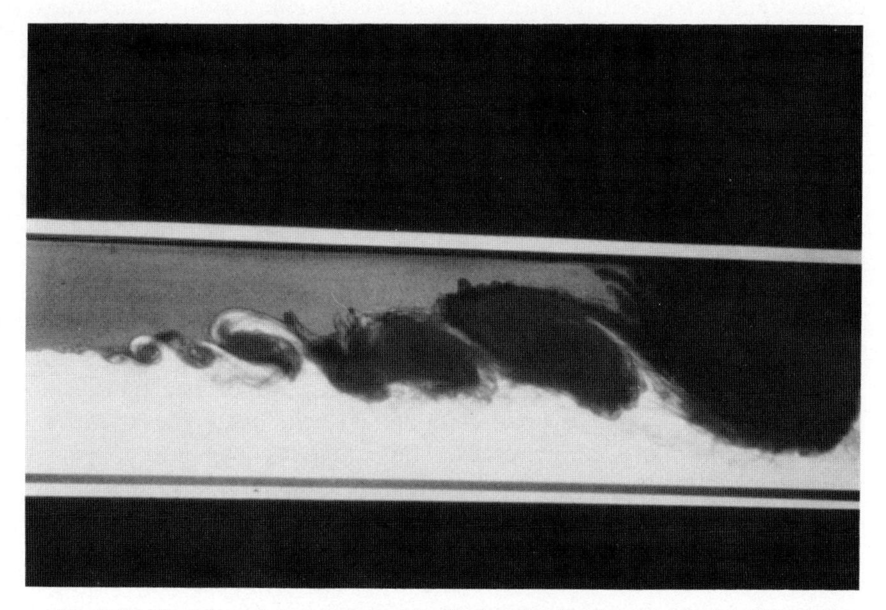

Fig. 2.4 Shear layer between two mixing flows, visualized by means of a pH-indicator reaction. Flow is from left to right. (From Breidenthal, 1981.)

(KI), sodium thiosulfate ($Na_2S_2O_3$), and starch. Iodine is produced by the following slow reaction:

$$Na_2S_4O_8 + 2KI \rightarrow Na_2SO_4 + K_2SO_4 + I_2.$$

Iodine reacts instantaneously with the thiosulfate and is reduced back to sodium iodide:

$$I_2 + 2Na_2S_2O_3 \rightarrow Na_2S_4O_6 + 2NaI.$$

When all the thiosulfate is used up, the free iodine reacts with starch and the solution turns blue immediately. The reaction time, that is, the time required for the mixture to turn blue, is therefore determined by the initial concentration of iodide, thiosulfate, and persulfate.

2.1.3. Smoke and Vapor Flow Visualization in Air Flows

2.1.3.1. General Properties of Smoke

The technique of marking filament lines or other flow contours in an air stream by means of smoke is, in principle, the same as visualizing the flow pattern of a liquid by the injection of dye. Although its application is not restricted to wind tunnels, smoke flow visualization now is a standard experimental tool in these facilities, and its technical advancement is closely related to the history of wind tunnels (Mueller, 1980). The idea of introducing smoke into an air stream for the purpose of visualization is, of course, old and had already been applied to scientific experiments by Ludwig Mach (1896), the son of Ernst Mach. An essential portion of the progress and refinement of the smoke technique is due to the work of Brown (1953) at the University of Notre Dame who systematically developed the generation of an appropriate smoke as well as the performance of suitable wind tunnels, later referred to as "smoke tunnels." Smoke flow visualization and its use in wind tunnels have been reviewed by Maltby and Keating (1962b) and by Mueller (1983).

The term "smoke" is not well defined, and it is used here in a wide sense, not only restricted to combustion products; we will also include steam, vapor, aerosols, mists, and tracer gases, which become visible without the need of applying particular optical techniques. The latter (i.e., a visible gas) is not a frequent case, and it might occur only in special situations when a tracer gas can be stimulated to emit a particular radiation (fluorescence). But, only a gas could fulfill the requirement of being a neutrally buoyant tracer, and in all the other cases that we describe by the general term "smoke" is the density of the tracer material orders of magnitude larger than the density of air. The solid or liquid particles of these materials normally have a diameter smaller, or even much smaller, than 1 μm; and this is why buoyancy effects, in a first-order approximation, become negligible. For, at the related low particle Reynolds number the drag force (in vertical direction) is very high ("Stokes law") and almost balances the force of gravity. In contrast to the methods described in Section 2.2., we are not interested here in the motion of single particles, and we rather consider the tracer material as one homogeneous, neutrally buoyant phase (i.e., a "smoke") whose motion becomes visible because it can scatter light by a considerable amount.

A great variety of smokes and methods to generate them have been described in literature. As in the case of dye one would like the smoke to fulfill a number of requirements (e.g., being nontoxic, neutrally buoyant, stable against mixing, and well visible). It turns out that these requirements are more difficult to meet than in the case of dye. All smoke is toxic to some degree, and one should prevent people in a laboratory from inhaling the smoke. The problem with buoyancy has already been mentioned. A general rule is that the mixing between smoke and air is more intense than between most of the dyes and water, so that one must provide much more care in stabilizing filaments or streamlines marked by smoke. No systematical investigation has been performed on the mixing of smoke lines with the flowing air, so that the differences in the various reported smokes consist either in the ease of their generation or in the quality of their visibility. The basic types of producing smoke are burning or smoldering tobacco, wood, or straw; vaporizing mineral oils; producing mist as the result of the reaction of various chemical substances; and condensing steam to form a visible fog.

2.1.3.2. Generation of Smoke

Cigarette smoke is a mixture of organic compounds and water, and it is often used if only small amounts of smoke are needed. The mean diameter of its particles is in the range between 0.2 and 0.4 μm (Kerker *et al.*, 1978). In order to scatter a sufficient amount of light to be readily seen, smoke particles should be larger than about 0.15 μm (Mueller, 1983). The interdependence of light scattering and particle size of combustion products has been investigated by Cashdollar *et al.* (1979). A smoke generator in which white pine wood is burned has been designed by Yu *et al.* (1972). The smoke is mixed with air from a compressed air reservoir and then filtered through steel wool.

From the point of view of laboratory safety it is more desirable to use vapors and mists rather than combustion products. The majority of smoke generators, some of them are commercially available, is based on the vaporization of hydrocarbon oils. Among these oils kerosene appears to be the best choice from the standpoint of mean particles size, light scattering characteristics, vaporization temperature, and flammability. A kerosene smoke (mist) generator mainly consists of a heating facility and a device in which the kerosene vapor is mixed with an air stream to form an appropriate mist. The Preston–Sweeting mist generator, originally operated with paraffin, is adequate for producing small quantities of smoke (Fig. 2.5; for details see Maltby and Keating, 1962b, or Warpinski *et al.*, 1972). A powerful generator that can produce large amounts of dense kerosene smoke has been designed by Brown at the University of Notre Dame and since then continuously improved (Mueller, 1983). Oil drips on the upper end of four electric strip heaters (Fig. 2.6) from where it evaporates. A blower forces the smoke to the smoke rake. An essential requirement is that the smoke, before entering the wind tunnel, is passed through a heat exchanger where it is cooled to room temperature or to the free-stream temperature in the tunnel. Finally, a filter should be applied, which removes larger particles and distributes the normal particles more uniformly in the smoke stream.

A frequently used chemical method for producing a dense white mist is based on the reaction of titanium tetrachloride, $TiCl_4$, with water. Hussain and Clark (1981) employ a smoke generator in which compressed air is pumped into a reservoir filled with liquid $TiCl_4$. Small droplets of $TiCl_4$ are entrained by the air stream, which is fed into a second reservoir filled

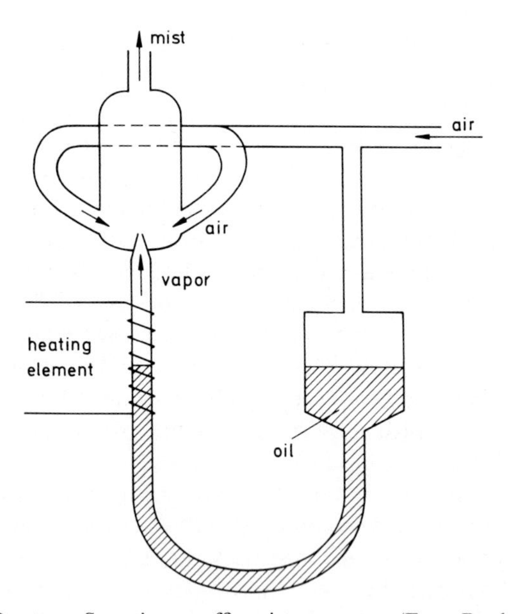

Fig. 2.5 Preston—Sweeting paraffin mist generator. (From Bradshaw, 1970.)

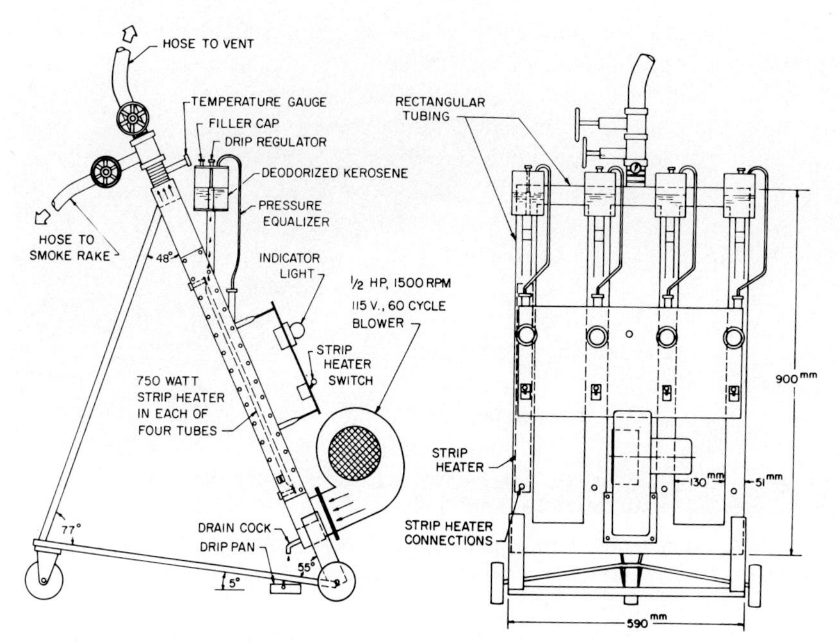

Fig. 2.6 Smoke generator designed at the University of Notre Dame. (From Mueller, 1983. Published by Hemisphere Publishing Corporation.)

with water. The result of the reaction is titanium dioxide, TiO_2, a white mist or smoke of good optical reflectivity. The TiO_2 smoke can also be generated by depositing a drop of $TiCl_4$ onto the surface of a test model in a wind tunnel. Upon reaction with the humidity in the air, a white stream of smoke will originate from the drop (Freymuth *et al.*, 1983). $TiCl_4$ has also been used in open-air large-scale experiments for measuring wind speeds (Fig. 2.7) (Dewey, 1971). A liquid particle aerosol generator for use in wind tunnels has been described by Griffin and Votaw (1973) and Griffin and Ramberg (1979). An aerosol of submicron-sized particles (mean diameter $d_p = 0.7\ \mu m$) is generated by bubbling compressed air through a container filled with the organic liquid DOP (di(2–ethylhexyl)–phthalate). The size range of the particles is controlled by a jet impactor from where the aerosol is fed through a flexible pipe system to the wind tunnel. The excellence of the visual results has been shown in many examples (e.g., Fig. 2.8).

The application of smoke visualization to the flow in recirculating wind tunnels creates the problem that after a certain time of operation the tunnel is completely filled with smoke. For this reason, and also since any smoke or aerosol is toxic to some extent, there has always been an inter-

Fig. 2.7 Visualization of the air flow initiated by an explosion shock wave by means of smoke tracers. The white-smoke tracers are seen against a black-smoke background. (From Dewey, 1971.)

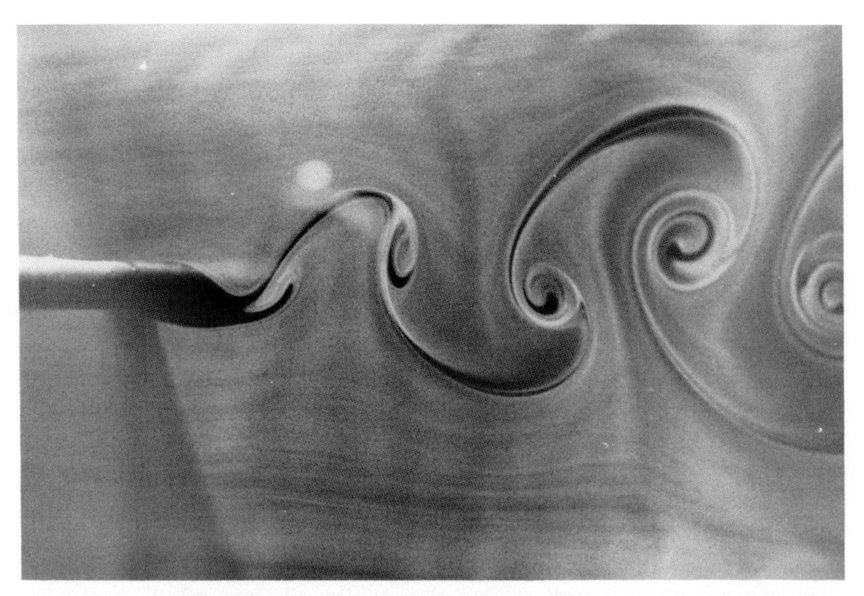

Fig. 2.8 Vortex shedding from a vibrating cylinder visualized by the introduction of a sheet of DOP aerosol into the center plane of the wind tunnel. (From Griffin and Ramberg, 1979. Published by Hemisphere Publishing Corporation.)

est in using steam as the visualizing tracer. The idea is that steam in combination with a cooling agent is introduced in the air flow where the mixture produces a visible fog (Prentice and Hurley, 1970). After heat exchange with the surrounding air, the fog will disappear, leaving the main air stream clean in the recirculating system, and perhaps, with the need of an air drying device. Bisplinghoff *et al.* (1976) and Parker and Brusse (1976) have developed such a steam-fog generating system in which liquid nitrogen serves as the cooling agent. Steam and liquid nitrogen are mixed turbulently in a mixing nozzle, and then expelled into the air stream through a pipe at the inlet of the wind tunnel's contraction section. The fog persists a given distance down the tunnel. Its optical appearance is claimed to compare with that of TiO_2, oil, or tobacco smoke, whose particles, however, are in the average smaller than the "particles" of water fog. The system has been demonstrated only in a relatively small wind tunnel. It requires an exact temperature control in order to produce a neutrally buoyant fog. An alternative cooling agent is dry ice (CO_2 pellets), which in contact with water, generates fog wanted for visualizing (Bouchez and Goldstein, 1975).

2.1.3.3. Introduction of Smoke into a Flow

For producing smoke lines in a wind tunnel, the smoke is released from a pipe or from a system of pipes ("rake") oriented parallel to the main air stream. Essential for the success of F.N.M. Brown's experiments was the fact that the smoke was released at the inlet of the contraction section so that the smoke lines were stabilized during the acceleration of the flow in this section. Special smoke tunnels with high contraction ratios have been developed (the tunnels at the University of Notre Dame have a ratio of 24 : 1!) in order to keep the turbulence level low and to prevent the smoke from an early mixing with the surrounding air. The flow from these straight indraft tunnels is exhausted into the outside air so that the interior of the laboratory is kept free from contact with possibly toxic materials.

Another way of introducing smoke in an air stream and producing relatively fine smoke lines with exact spatial and temporal control is the smoke wire. Apparently this technique has been developed simultaneously and independently in different laboratories (Yamada, 1973, 1979; Corke *et al.* 1977; Nagib, 1979; Torii, 1979). Instead of an external smoke production the smoke, which is here oil fog, is generated inside the wind stream by evaporating oil from an electrically heated wire. The wire, usually made of stainless steel, approximately 0.1 mm in diameter, is oriented vertically or horizontally and normal to the main flow. It can be mounted on a portable probe or span the whole cross section of a wind tunnel (Fig. 2.9). Before running the experiment the wire is coated with a

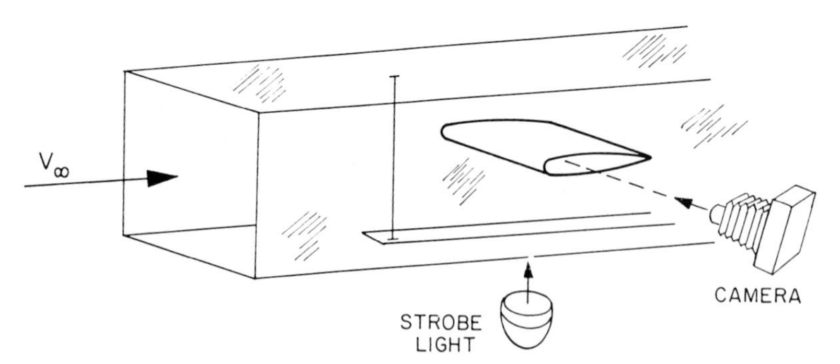

Fig. 2.9 Locations of smoke wire, camera, and light source for vertical wire position in wind tunnel. (From Mueller, 1983. Published by Hemisphere Publishing Corporation.)

very thin sheet of oil (kerosene, paraffin). Batill and Mueller (1981) suggest the application of model train smoke; and Sieverding and van den Bosche (1983) report on the generation of colored smoke if the wire is coated with a paste of dye and oil. The coating is done either manually by wiping the wire with a fine, small brush or a cotton-tipped applicator, or a drop of oil is forced down the wire by applying pressure to a small oil reservoir (Corke *et al.*, 1977). After the wire is coated with oil, small oil beads or minute droplets form, and at each of the beads a smoke filament originates when the wire is heated. The spacing between these lines is influenced by the wire size and the oil surface tension. The heating of the wire, usually done by applying dc or ac voltage of about 50 V, must be synchronized with the illumination system and the camera. Figure 2.10 is a schematic block diagram of such system in which the strobe light and the wire are initiated by pressing the camera button. Since the smoke wire allows for generating fine sheets of controllable smoke lines (Fig. 2.11), it has already found a wide application (see, e.g., Mueller and Batill, 1982; Sakamoto and Arie, 1983; Leonard *et al.*, 1983; Wei and Sato, 1984).

Both the smoke injection pipe and the smoke wire are mechanical disturbances in the flow to be investigated, though the interference of the very thin wire with the air flow might be negligibly small. It would be advantageous if the smoke could be generated in the flow without introducing a mechanical device. The deposition of a drop of $TiCl_4$ on a solid surface (Freymuth *et al.*, 1983; see preceding section) is such a solution of the problem, though it does not provide a time control to the smoke generation. Such a control is realized in the technique communicated by Farukhi (1983). The surface close to which the flow should be visualized is pasted with a substance (ammonium chloride mixed with a black dye) that sublimates when irradiated by a laser beam. The application of a 5 W

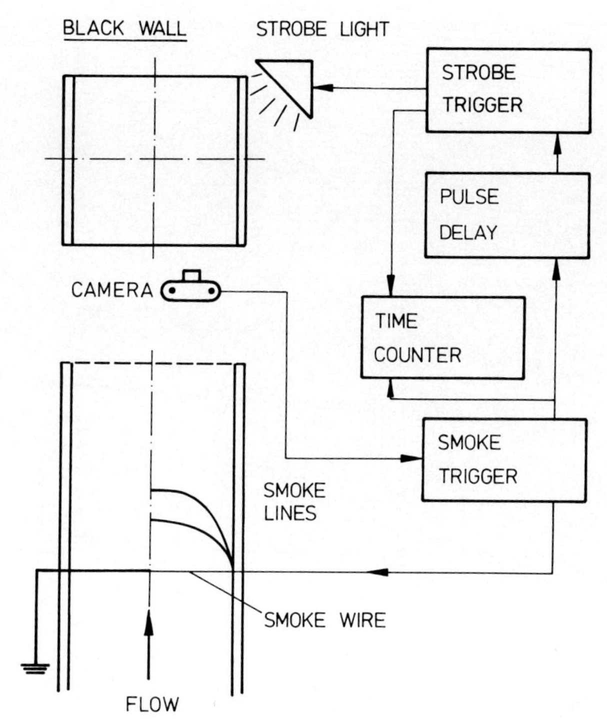

Fig. 2.10 Synchronization of smoke wire with camera and illuminating strobe light. Triggering initiated from camera. (From Torii, 1979. Published by Hemisphere Publishing Corporation.)

argon laser allows for generating the smoke at the desired moment. In principle the same is the stimulation of fluorescence radiation from a tracer gas in the gas flow (Cenkner and Driscoll, 1982). Helium as the test gas was seeded with iodine vapor. The tracer gas can be excited by means of the green light from an argon laser, and the molecules in the excited gas volume rapidly decay to lower energy levels, thereby emitting yellow and red fluorescing light. The size of the exciting laser beam permits a high spatial control of the fluid volume to be marked. Though this technique cannot be considered a smoke method, its application delivers results comparable to the classical smoke visualization.

In the beginning of this section it was pointed out that the term "smoke" is not well defined. If we take "smoke" just as a tracer material that can be seen in gas flows, we would as well include visible tracer gases, provided that such visible gases exist. Gases are invisible under normal conditions. However, in Section 3.1 methods will be discussed, which enable us to visualize the mixing of gases with different optical

Fig. 2.11 Smoke lines around an airfoil in low Reynolds number wind-tunnel flow. (From Mueller and Batill, 1981.)

properties. Therefore, it is possible to use tracers even in gaseous form (e.g., ozone) if the respective optical visualization methods are applied. The results, again, compare to classical smoke visualization (see Section 3.1).

2.1.3.4. Illumination

Illumination arrangements depend on what is aimed to become visualized. The observation of smoke lines needs illumination from the front; records should be taken in a direction in which the scattering characteristics of the smoke particles have a maximum. Conventional illumination is appropriate for this purpose, e.g., mercury lamps, halogen lamps, and spot lights. If the smoke fills a larger volume in the flow field, for example, a separated flow area, a boundary layer or a wake, one often is interested in illuminating only a plane sheet so that certain flow structures become visible. Although such light sheets can be produced with conventional lamps (Cherdron *et al,* 1978), the best way is to expand the beam from a powerful laser by means of a cylindric lens in one plane (see, e.g., Philbert *et al.,* 1979; Hussain and Clark, 1981; Véret, 1985). The thickness of such a sheet can be made smaller than 1 mm (Fig. 2.12); observation of

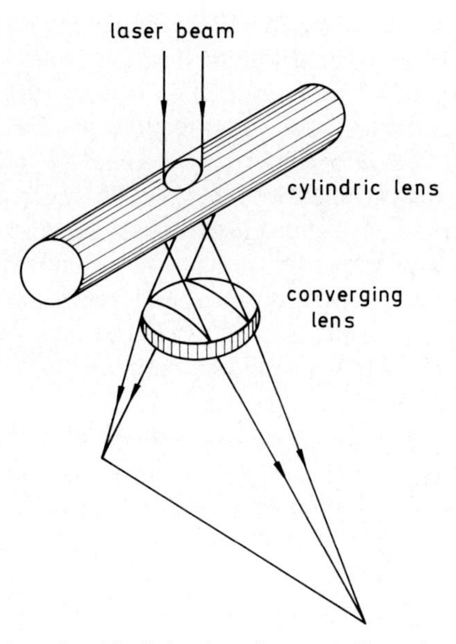

Fig. 2.12 Generation of a thin light sheet by expanding a laser beam in one plane. (From Philbert *et al.*, 1979.)

the flow and its structures occurs normal to the illuminated plane or under a certain angle to it.

A thin, light sheet, normal to the main flow direction, is the appropriate illumination for recognizing vortical structures in flows that are totally seeded with smoke. This technique has been applied to channel flow (Sugiyama *et al.*, 1983), to numerous wind tunnel experiments, and even, in a modified way, for the purpose of visualizing the trailing vortices left behind by an aircraft on a runway after take-off (Hallock *et al.* 1977). Special wind tunnels have been developed, either filled with smoke (Maltby and Keating, 1962b) or with vapor (McGregor, 1961), in which a plane light sheet normal to the tunnel axis illuminates the large-scale vortices produced by respective test models (see Section 2.1.4).

2.1.3.5. *Applications*

The applications of smoke visualization are numerous, in any possible velocity range, in wind tunnels and other flow facilities, in laboratory and open air experiments. Only a few characteristic examples can be mentioned. The development of low-drag road-vehicle shapes due to the improvement of their aerodynamics has made great progress, since the air flow around full-scale models has been visualized by means of smoke

(Fig. 2.13) (Hucho and Janssen, 1979; Imaizumi *et al.*, 1979; Ahmed, 1983). The stability and smoothness of smoke lines depends to a great portion on the degree of free turbulence in the wind tunnel, and it has often been criticized that good-looking flow pictures like Fig. 2.13 are taken at too low velocities in order to keep the associated Reynolds number as well as the turbulence level low. However, free turbulence and model Reynolds number are not directly related, and there exists practically no upper limit of speed for smoke-line visualization, if the tunnel is provided with a severe reduction ratio and heavy antiturbulence screening. Smoke visualization has been extended even to the supersonic range (Goddard *et al.*, 1959; Mueller and Goddard, 1979; Roberts and Slovisky, 1979; Batill *et al.*, 1982).

The contours of vortices, wakes, and other separated flows can be visualized clearly by means of smoke (Fig. 2.14) (see, e.g., Baker, 1979; Freymuth *et al.*, 1983; Mueller and Batill, 1982; Sadeh and Brauer, 1980; Sakamoto and Arie, 1983; Taneda, 1978; Zdravkovich, 1968, 1973). There is only little mass exchange between the outer flow and the vortex or separated flow regime, which in the experiments usually appears filled with smoke. Coherent structures may become visible in the free shear

Fig. 2.13 Smoke lines around a road vehicle in a full-scale wind tunnel. (Courtesy of Volkswagenwerk AG.)

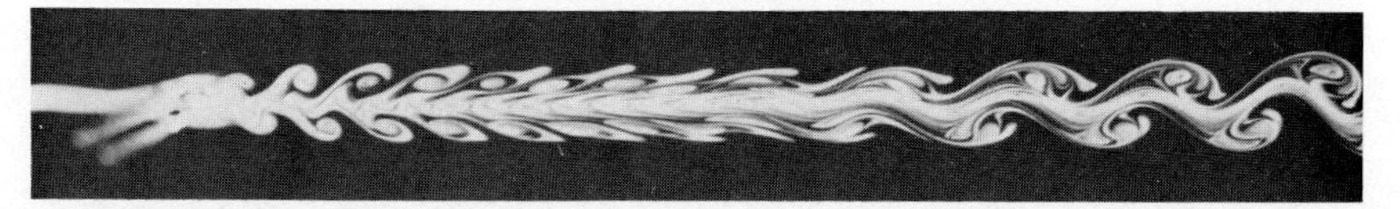

Fig. 2.14 Vortex systems in the wake of a group of three cylinders. Smoke photograph taken at a Reynolds number Re $\cong$ 100. (From Zdravkovich, 1968.)

layers of wakes and free jets (Hussain and Clark, 1981; Mensing and Fiedler, 1980; Michalke, 1964; Perry *et al.*, 1980; Zaman and Hussain, 1980). The experiments of Head and Bandyopadhyay (1981) have permitted deep insight into the vortical structure of turbulent boundary layers. Vortex formation in convective flows have been visualized by Inaba (1984), Kimura and Bejan (1983), Linthorst *et al.* (1981), Morrison and Tran (1978), or Ruth *et al.* (1980).

Dewey and Walker (1973) have shown that quantitative results on velocity, pressure, and density can be derived from the observation of smoke puffs, which have been released through small holes in the plane side wall of a shock tube. If one could measure the instantaneous concentration of smoke in an air flow, one should be able to derive quantitative values of the mass transfer and other related fluid mechanical parameters (e.g., turbulence characteristics). The development of such methods involves in any case the physics of Mie scattering from the smoke particles. A number of different attempts in that field (e.g., Becker *et al.*, 1967; Gad-el-Hak and Morton, 1979; Long *et al.*, 1980) are beyond what we mean with visualization.

2.1.4. Screen Methods

It has already been mentioned that light sheets as reviewed by Véret (1985) can be used to illuminate a plane normal to the mean flow direction in special wind tunnels which are, to a certain degree, filled with either smoke or vapor. The technique is then called smoke screen or vapor screen, respectively. It allows for the visualization of the systems of trailing vortices, which separate from test models. A schematic view of the test section, light source, vapor (or smoke) screen and recording camera is shown in Fig. 2.15.

While the application of smoke tunnels is restricted to subsonic flow (Maltby and Keating, 1962b), vapor screens are usually produced to observe the flow in supersonic wind tunnels. The principle then is to run the tunnel with moist air; the air cools as it expands through the supersonic

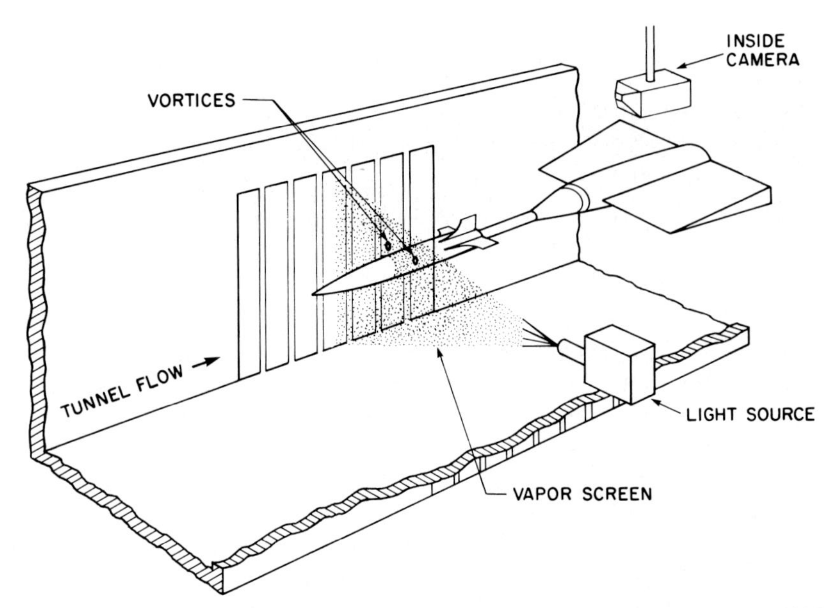

Fig. 2.15 Perspective view of vapor screen method. (From Snow and Morris, 1984.)

nozzle, and the moisture condenses to form a fog in the test section. The uniform distribution of fog in a cross section normal to the tunnel axis is disturbed due to the wake from a test model. McGregor (1961) has investigated the rate of humidity required to produce a satisfactory screen at different Mach numbers (up to $M = 2$) and as a function of tunnel pressure and temperature. More recently, Snow and Morris (1984) systematically investigated the operational conditions of the vapor screen. They also describe a periscope system for observation so that the recording camera can be removed from the inside of the tunnel.

An alternate way for examining the flow pattern in a plane normal to the mean stream is a grid of wires and short tufts attached or glued in the nodal points of the grid. Again, such "tuft screens" have been used to visualize the vortex shedding behind various model bodies (Bird, 1952; Gersten, 1956; Taneda, 1978), or the interaction regime of an air jet and a cross wind (McMahon *et al.*, 1971). The grid is placed in the wind tunnel (subsonic!) normal to the mean flow direction, and the tuft pattern is observed or photographed from downstream (Fig. 2.16). The resulting pattern is similar to the results obtained with the smoke-screen or vapor-screen technique. But beyond a qualitative visualization of the flow pat-

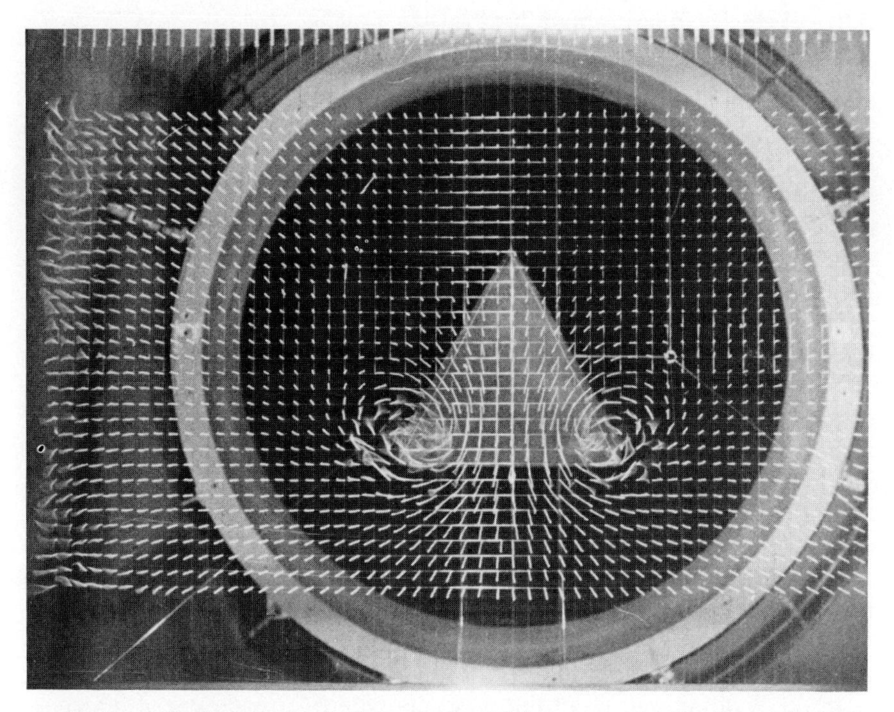

Fig. 2.16 Trailing vortices behind an inclined delta-wing as visualized by a tuft screen. (From Gersten, 1956.)

tern one may well perform quantitative measurements of the velocity components in the plane normal to the mean flow. If a homogeneous air stream of constant velocity is directed normal to the grid, the orientation of all tufts will be perpendicular to the screen. A disturbance due to the vortex or a cross flow causes the tufts to be deflected from their original, parallel orientation. The vertical projection of the new tuft positions onto the plane of the grid yields the respective velocity components (normal to the mean direction). However, the recording of the tuft pattern with a camera may introduce a severe error source. Since the camera is usually positioned within a small distance downstream of the grid, one obtains a central rather than a vertical projection of the screen on the photograph. The tufts in the outer zone of the screen therefore appear deflected even in the undisturbed case of a homogeneous flow, and these errors must be corrected for a quantitative evaluation.

2.2. Velocity Measurement by Particle Tracing

In this section we consider the case that single foreign particles are added to a flowing fluid and move along with the flow. We assume that the concentration of particles in the fluid is so low that we can distinguish the movement of the individual particles, and that we may measure, by means of an appropriate observation and recording system, the velocity of one single particle. This is a way, for example, to measure the wind speed by means of tracking the motion of small balloons in the air. Measuring the wind speed in this example means that one assumes the movement of the balloon and the air, or of the particle and the fluid, to be identical. This is not necessarily the case. The velocity of the tracer particle might deviate in direction and magnitude from the fluid velocity. This leads us to the basic question for all flow measurements using tracer particles: Does the particle follow the flow? The general answer is "no." But in some cases, the difference between particle and flow speed can be very small, and one might obtain very reasonable results. We shall therefore first study the motion of a single particle, which has been released to a fluid flow. From this we may obtain an idea on the accuracy of such flow velocity measurements. Subsequent to this flow analysis, the generation, properties, and use of particles for certain flow cases is reported, followed by a review of such observation and recording systems, which provide visual information of the velocity field (whole-field methods).

Analyzing the motion of a foreign particle in a fluid flow is an essential problem also for a number of related fields, particularly for the methods of optical probe velocimetry depending on the light scattered from tracer particles, like laser–Doppler anemometer or laser–2-focus anemometer, and for the fluid mechanics of dispersed systems or multiphase flow. Therefore, further and more detailed information is available in the review literature of these fields [e.g., Clift *et al.* (1978), Durst *et al.* (1981), Somerscales (1981)].

2.2.1. Motion of a Single Particle in a Fluid Flow

The motion of particle-fluid systems has been the subject of numerous theoretical and experimental investigations. For the present purpose we assume that the concentration of particles in the fluid phase is so low and the particle size so small, that the flow of the fluid phase (e.g., its velocity field and turbulence characteristics) is not disturbed by the presence of the particles. The particle-free flow is described by the Navier–Stokes equations. Heat transfer to the particles and particle interactions are neglected. A particle released to the fluid is subjected to forces exerted by the fluid flow, and eventually to certain external forces. The question whether the particle follows the flow will be discussed for a very simple case: We shall investigate the velocity and trajectory of a small particle, which is released to a plane (two-dimensional) shear flow.

The components of the fluid velocity in the x–y plane are assumed to be $u_F = u_F(y)$ and $v_F = 0$. The particle of mass m_p is accelerated in x direction due to the drag force, and the x component of its equation of motion is

$$m_p \frac{du_p}{dt} = c_D \frac{\rho_F}{2} A_p(u_F - u_p)\sqrt{(u_F - u_p)^2 + v_p^2}, \tag{2.1}$$

where it has been assumed that the particle has two velocity components, u_p and v_p, A_p is the cross section of the particle, and c_D is a drag coefficient. No external forces, such as gravity or buoyancy, are included in Eq. (2.1). The sole force, which determines the motion of the particle in x direction is the drag force, which arises from the velocity difference between the particle and the fluid. Equation (2.1) applies, in general, to all particle shapes if one uses the respective values of c_D and A_p. However, little is known about drag coefficients for particles other than spheres, and we must therefore restrict ourselves to spherical particles. In this case the particle mass is $m_p = \frac{1}{6}\pi d_p^3 \rho_p$, with d_p being the diameter and ρ_p the density of the particle, and $A_p = \frac{1}{4}\pi d_p^2$.

In order to solve Eq. (2.1) for u_p, v_p (the second component of the equation of motion is discussed later), the fluid velocity u_F must be determined from the Navier–Stokes equation. The drag coefficient c_D is a function of the velocity difference $\sqrt{(u_F - u_p)^2 + v_p^2}$ or of the respective particle Reynolds number

$$c_D(\mathrm{Re}_p) = c_D\left(\frac{d_p\sqrt{(u_F - u_p)^2 + v_p^2}}{\nu_F}\right),$$

where ν_F is the kinematic viscosity of the fluid. The dependence of c_D on Re_p is known for spheres in form of an empirical relationship for a wide range of Reynolds number. Since tracer particles should be small, the respective particle Reynolds numbers are low, and in many cases it is sufficient to use the value according to Stokes's law, $c_D = 24/\mathrm{Re}_p$, which is valid for $\mathrm{Re}_p \leq 1$. A number of interpolation formulas are known, which apply at higher values of Re_p, e.g., Morsi and Alexander (1972), describe the drag coefficient of a sphere as a power series of $(1/\mathrm{Re}_p)$:

$$c_D = K_0 + K_1(1/\mathrm{Re}_p) + K_2(1/\mathrm{Re}_p)^2 + \cdots.$$

The constants K_i must be determined from the empirical curve. During the acceleration of a particle, the drag coefficient changes according to the instantaneous value of the velocity difference $(u_F - u_p)$ or the particle Reynolds number. Such a change can be accounted for in a numerical calculation of the particle's velocity and path. However, this is not an exact procedure because the usual c_D values have always been measured under stationary flow conditions, and very little is known about the true dependence of c_D on time.

If Stokes' law applies and if the y component of the particle velocity vanishes, the equation of motion, Eq. (2.1), for the spherical particle, reduces to

$$\frac{du_p}{dt} = K(u_F - u_p), \qquad K = \frac{18\nu_F}{d_p^2}\frac{\rho_F}{\rho_p}. \tag{2.2}$$

Equation (2.2) has a very simple solution if u_F is taken as constant. With the initial condition $u_p = u_{p0}$ for $t = t_0$ one has

$$u_p = u_F[1 - e^{-K(t-t_0)}] + u_{p0}e^{-K(t-t_0)}, \tag{2.3}$$

and if the particle is released to the flow at time $t = 0$ with zero velocity:

$$u_p = u_F(1 - e^{-Kt}). \tag{2.4}$$

The particle velocity approaches exponentially the fluid speed. The response of the particle is the faster, the smaller the particle size and the density ratio (ρ_p/ρ_F). The equation does not hold if the ratio (ρ_p/ρ_F) is smaller than one, for example, in the case of a gas bubble in a water flow. With the many simplifications involved in Eq. (2.4), this equation cannot describe a realistic situation, and it can only qualitatively explain the possible lag between particle speed and true flow velocity. In particular, it is not possible to draw any conclusion on whether a particle can follow the velocity fluctuations in a turbulent flow. If the particle Reynolds number, Re_p, exceeds the value of 1, the particle's approach to the fluid velocity will be more rapid than Eq. (2.4) would predict (Chen and Emrich, 1963).

Equation (2.1) is only one component of the equation of motion. The y component is taken in the vertical direction, and therefore has to account also for the force of gravity. Since a plane shear flow with a velocity profile $u(y)$ has been assumed, an additional "lift" force acts upon the particle in the y direction. This force is due to the existence of a velocity gradient (du/dy) and has been analyzed by Saffman (1965). It is explained by a pressure difference between the top and the bottom of the particle in the shear flow. The direction of the "lift" force is in the direction of the velocity gradient, and therefore, it causes the particle to move across streamlines of the fluid flow. Considering the force of gravity, the drag force, and the "lift" force, the y component of the equation of motion becomes

$$m_p \frac{dv_p}{dt} = -gm_p - c_D \frac{1}{2} \rho_F A_p v_p \sqrt{(u_F - u_p)^2 + v_p^2}$$

$$+ \mathbf{a} \frac{D_p^2}{4} \rho_F (u_F - u_p) \sqrt{\nu_F |du_F/dy|}. \qquad (2.5)$$

This form only applies to a spherical particle, because the "lift" force has been derived by Saffman for a sphere. According to Halow and Wills (1970), the constant **a** can be taken as 32.2. In Eq. (2.5), g is the acceleration of gravity, and the rest of the notation is the same as for Eq. (2.1).

The two equations of motion, Eqs. (2.1) and (2.5), are coupled through u_p, v_p. It is not possible to derive for Eq. (2.5) such a simplified closed solution like Eq. (2.3) or Eq. (2.4) for Eq. (2.1). The two equations are used here for explaining qualitatively the two major sources for deviations of the particle velocity from the fluid velocity: a velocity lag or a deviation in the amount of velocity is due to the inertia of the particle; a deviation in the direction is caused by Saffman's lift force existing in a flow with a velocity gradient (shear flow). One has to be aware of these two possible errors if it is aimed at measuring a flow velocity by means of tracer particles.

The lift force can be of great influence on the motion of particles, which are entrained into the boundary layer close to a solid wall. The velocity gradient present in such a layer generates a force, which is directed away from the wall. As a result, a particle-free region develops close to the wall (Cox and Mason, 1971). The thinner the boundary layer, the greater the velocity gradient. For the very thin boundary layer behind an unsteady shock wave in air, Merzkirch and Bracht (1978) have shown that the acceleration of a particle away from the wall (i.e., normal to the mean flow direction) can exceed the acceleration of gravity, g, by several orders of magnitude.

Another example of a flow with velocity gradient is a vortex. In a free vortex, whose velocity distribution resembles that of a potential vortex, or in any rotating flow other than a solid rotation, a tracer particle is subjected to two forces, which act in radial direction and cause the particle path to deviate from the true flow direction: one is the centrifugal force, and the other is equivalent to the mentioned lift force and is due to the radial gradient of the tangential velocity of the vortex. The velocity in the potential vortex is such that the velocity approaches zero at the outer edge of the vortex and has a maximum near the center; the associated "lift" force is therefore directed radially toward the center. Small particles added to such a steady vortex flow accumulate at a radial position where these two forces balance each other (Fig. 2.17). If one uses neutrally buoyant particles, which have the same density as the fluid, the disturbing effect of the centrifugal forces is eliminated. However, the "lift" force is still present, and one observes, therefore, a drift of such particles toward the center of the vortex (Hasinger, 1968). It is also obvious, that this lift force is not present in a flow, which rotates like a solid body.

The equation of motion, Eq. (2.1), applies, in principle, also to compressible fluid flows, if one uses an appropriate expression for the drag coefficient. For this type of flow, the drag coefficient is a function of body Reynolds number and Mach number. From a correlation of empirical data, Crowe (1967) has derived expressions, which describe the c_D values of spherical particles in the Mach-number range up to $M = 2$ and for Reynolds numbers between 10^{-1} and 10^3. A characteristic feature of compressible flows is the formation of shock waves, which represent a surface at which the flow properties change discontinuously. Particles of finite mass and size cannot follow such an abrupt change of the state of motion. In traversing such a discontinuity surface in a supersonic gas flow, a tracer particle needs a certain relaxation time until its state of motion has been compensated by that of the flowing gas. For determining these relaxation times (e.g., Rudinger, 1964), one must not only consider the intertia forces, but also account for the heat transfer from the shock-heated gas to the particle. Measurements of such velocity-relaxation times of submicron particles accelerated by a shock wave have been performed by vom Stein and Pfeifer (1972).

Fig. 2.17 Heavy particles (specific gravity = ~2) stabilized in a free-water vortex; average particle diameter = 30 μm. (From Hasinger, 1968.)

The motion of tracer particles can, of course, be influenced by a number of additional types of forces, particularly by such not affecting the fluid molecules and so causing discrepancy between the two movements. Just to mention one that creates a complication for heat transfer studies: Bez and Frohn (1982) identified a ''thermophoretic'' force acting on tracer particles in convective flows where a temperature gradient is present. This force is proportional to the temperature gradient and directed away from a hot surface so that a particle-free region near the hot body can develop.

The preceding theoretical considerations cannot provide information on what tracer particles do in a turbulent flow. The fundamental question here is: Up to which frequency can the particle follow the fluctuations of a fluid element? This problem requires a different approach, and the adequate form of the equation of motion for investigating this case is the Bassett–Boussinesq–Oseen (BBO) equation. Detailed discussions of this equation and its different terms as well as possible simplifications can be found in the book of Hinze (1975), but also in the pertinent literature on laser–Doppler anemometry [e.g., Durst *et al.* (1981), Somerscales (1981)]. On the basis of the BBO equation, Hjemfeldt and Mockros (1966) have determined limiting frequencies up to which specific particles can be regarded to follow flow fluctuations of given fluids with sufficient accuracy. This information is of extremely great relevance for optical probe methods where these fluctuations can be measured in one point of the flow (see Section 2.1). But whole-field methods, and, again, this is what we mean by ''flow visualization,'' can barely resolve the time history in such detail, and this kind of analysis of the motion of tracer particles (i.e., their behavior with respect to turbulent fluctuations) will not be discussed further.

2.2.2. Selection of Particles

Tracer particles should meet a number of general requirements, which have already been mentioned in connection with materials used as dyes or smoke. What remains are two basic demands, that of high visibility, and that of minimizing the slip between particle and fluid flow. These requirements are somewhat contradictory because the visibility of the particle increases with its size; whereas, the particle follows the flow better the smaller its size. The second requirement implies also that the particle be neutrally buoyant. An exception are particles, which float or swim at the free surface of a water flow where their motion is measured.

Surface flow patterns in a water channel have been visualized in the early experiments by Prandtl, Tietjens, and co-workers in Göttingen who so established the existence of a wall boundary layer and the formation of free vortices behind obstacles. Aluminum flakes had been used as the floating tracer particles. Lycopodium powder, wood sawdust, or pulverized polyesters are further materials suitable for the visualization and measurement of flow at a free water surface (Clayton and Massey, 1967; Dimotakis *et al.*, 1981; Brandone and Bernard, 1971). The motion of small aluminum flakes at the liquid surface is affected by surface tension, and the recorded particle motion does not necessarily indicate how the flow behaves in the interior of the fluid below the surface. Douglas *et al.* (1972) have observed situations in which the motion of particles suspended in a liquid indicated a fluctuating flow pattern below the surface, while other particles floating at the surface did not follow these fluctuations. This is due to the damping effect of the surface tension.

The sedimentation velocity of a spherical particle of diameter d_p and material density ρ_p, in a fluid of density ρ_F and kinematic viscosity ν_F, assuming that Stokes' law applies, is given by

$$v_s = (gd_p^2/18\nu_F)\{(\rho_p/\rho_F) - 1\}, \tag{2.6}$$

where g is the gravitational acceleration. A particle is neutrally buoyant if $v_s \simeq 0$. Since the condition $\rho_p/\rho_F \simeq 1$ cannot always be fulfilled, neutral buoyancy can also be approached by using extremely small (i.e., micron-sized particles). This applies particularly if the fluid is a gas, usually air. If one requires the particle to sediment during the time period of an experiment, Δt_{exp}, by not more than one particle diameter, then the appropriate size of the particle is determined by

$$d_p \leq \frac{18\nu_F}{g\,\Delta t_{exp}\{(\rho_p/\rho_F) - 1\}}. \tag{2.7}$$

The required particle diameter is inversely proportional to the intended time of observation. The larger particles have a greater sedimentation velocity than the smaller ones. In many cases a liquid tracer material is sprayed by means of an atomizer into the test gas where it is suspended in the form of small droplets of different size. If the droplets are sprayed before the flow is started (i.e., if the gas is at rest) the larger particles will settle on the floor of the flow facility after a certain time. One may therefore wait a definite time before starting the experiment so that all particles with a diameter greater than a predetermined value settle on the floor, and only particles with a diameter equal to or smaller than the desired value are suspended in the gas. This procedure requires a device to start and establish the flow within a very short period, as in a shock tube (Chen and Emrich, 1963).

Matching the density of the tracer material with that of a liquid fluid can be achieved by the proper selection of the two materials. Mixing water with glycerol delivers a liquid solution with $\rho_F < 1$ g/cm^3, and certain liquids are available with densities up to $\rho_F = 1.6$ g/cm^3 at room temperature, for example, freon with $\rho_F = 1.62$ g/cm^3 at 30°C, or carbon-tetrachloride with $\rho_F = 1.59$ at 25°C. The question is whether the viscosity and other properties of such liquids can be tolerated. Since neutral buoyancy is relatively easy to arrange for liquid flows, the respective tracer particles then do not need to be extremely small, which is favorable for the intensity of the scattered light. Tracer particles with diameters of approximately 100 μm are often used under these circumstances.

For most of the equations used in this section it has been anticipated that the tracer particle has a spherical form. This is indeed the case for most of the particles used for flow tracing experiments. The light scattering properties of spheres are well known. A number of experimenters, however, prefer to use nonspherical, anisotropic particles and claim that these tracers align with the flow, which appears to be questionable. A possible advantage of this case might be that the light be scattered in certain directions with a higher degree of intensity.

A number of possible tracer particles and their respective application are listed in Table 2.3. The reader's attention should also be directed to the pertinent literature on laser–Doppler velocimetry where materials suitable for the production of tracer particles are listed. Very frequently used in water flows are Pliolite particles; this is the commercial name of polyvinyltoluene butadiene, a solid white resin, which can be ground to yield particles usually between 10 μm and 200 μm in diameter. Pliolite particles are insoluble in water, exhibit a high degree of reflectivity, and with a density of 1.02 g/cm^3 they are suspended virtually indefinitely in water (Telionis and Koromilas, 1978; Carey and Gebhart, 1982). Polystyrene is another material very often applied for generating tracers. Small spherical beads of polystyrene can be manufactured with a diameter of about 0.1 mm; the density of these beads slightly exceeds the density of water. The density of polystyrene beads can be manipulated by the treatment with acetone or benzene, and it is thus possible to produce neutrally buoyant particles for liquids like solutions of glycerol in water (Gent and Leach, 1976). With the application of a luminescent paint to polystyrene one may produce tracer particles of bright reflectance (Nakatani and Yamada, 1982). Powe *et al.* (1980) report on the direct generation of tracers from a fluorescent paint for use in the flow of silicone oil. Other chemicals from which white spherical beads of about 20 μm in diameter can be generated have been used by Haselton and Scherer (1982).

Aluminum and Magnesium flakes, 20 to 50-μm long and about 5-μm

TABLE 2.3

Tracer Particle/Fluid Combinations Used for Velocity Measurements

Working fluid	Particle	Diameter	Reference
Water	Pliolite	40–200 μm	Carey and Gebhart (1982), Chiou and Gordon (1976), Kao and Pao (1980), Koromilas and Telionis (1980), Mezaris *et al.* (1982), Nychas *et al.* (1973), Telionis and Koromilas (1978)
Water and water/glycerol	Polystyrene	10–200 μm	de Verdiere (1979), Douglas *et al.* (1972), Gent and Leach (1976), Greenway and Wood (1973), Nakatani and Yamada (1982), Seki *et al.* (1978)
Water	Hollow glass spheres	25 μm	Kao and Kenning (1972), Pan and Acrivos (1967)
Water/glycerin, silicone oil	Bees wax	0.2–1.0 mm	Mallison *et al.* (1981), Maxworthy (1979)
Water	Aluminum and magnesium flakes	10–100 μm	Coutanceau and Bouard (1977), Ozoe *et al.* (1981), Gau and Viskanta (1983)
Water, CCl_4	Kalliroscope flakes	30 μm	Matisse and Gorman (1984), Rhee *et al.* (1984)
CCl_4	Organic dye particles		Trinh *et al.* (1982)
Water	Rayon flocs	150–800 μm	Hoyt and Taylor (1982)
Water	Mixed droplets	20–200 μm	Charwat (1977a), Yin *et al.* (1973)
Air	Talc	10 μm	Sparks and Ezekiel (1977)
Air	Lycopodium	30 μm	Chen and Emrich (1963), Weinert *et al.* (1980)
Air	Glass spheres	20 μm	Philbert and Boutier (1972), Klein *et al.* (1980)
Air	Oil droplets	1 μm	Emrich (1983)
Air	Helium bubbles	1 mm	Colladay and Russel (1976), Kent and Eaton (1982)

thick, are nonspherical tracers, which have been used since the very early days of experimental hydrodynamics. With the appropriate illumination these particles generate excellent flow pictures as documented in many experiments of M. Coutanceau (Fig. 2.18; see e.g., Coutanceau and Bouard, 1977; Coutanceau and Thizon, 1981). Other nonspherical solid tracers for liquid flows have been described by Matisse and Gorman (1984) and Rhee *et al.* (1984). These tracers are platelike crystalline flakes

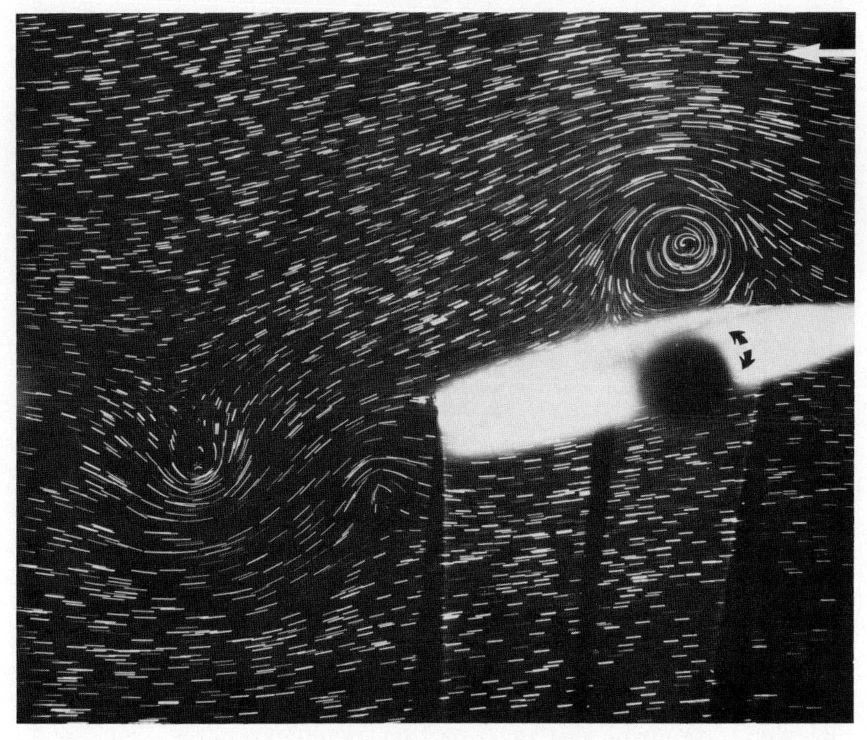

Fig. 2.18 Unsteady flow around an elliptic body oscillating at a frequency of 0.17 Hz. The fluid is water and the tracer particles are lycopodium; Reynolds number = 3300. (Courtesy M. Coutanceau, University of Poitiers, France.)

(trademark: Kalliroscope flakes), with dimension of about 5 μm $\times$ 30 μm $\times$ 0.1 μm, and a density of $\rho_p = 1.62$ g/cm^3. They are neutrally buoyant in freon ($\rho_F = 1.62$ g/cm^3 at 30°C) or carbon tetrachloride ($\rho_F = 1.59$ g/cm^3 at 25°C), have a high refractive index ($n = 1.85$). Particles of this geometrical shape are supposed to align along the axis of principal normal strain (see also Carlson *et al.*, 1982).

Charwat (1977a) describes the generation of neutrally buoyant droplets for use as tracers in a water tunnel. The droplets of size 20–200 μm in diameter are produced by mixing red mineral oil ($\rho_p = 0.827$ g/cm^3) and carbon tetrachloride ($\rho_p = 1.59$) such that the resulting fluid matches the density of water in which the droplets are immiscible.

Neutrally buoyant particles for air flows are hardly available, and the general tendency is therefore to use much smaller particles, solid or liquid, for reasons that we have discussed in connection with the sedimentation velocity [Eq. (2.6)]. An exception in this situation are soap bubbles filled with helium, which can be made neutrally buoyant in air. A bubble generator, which is commercially available consists of a head where the bubbles are formed (Fig. 2.19) and a console where the flow of helium, soap solution, and air is controlled. Bubbles can be generated with a diameter of about 1 mm and a rate of up to 300 bubbles per second. The user has to tolerate that the bubbles hitting the wall of models and test section are wetting these surfaces after some while. An application of these helium bubbles is reported by Colladay and Russell (1976) and by Kent and Eaton (1982).

2.2.3. *Illumination and Recording Systems*

The motion of tracer particles in a flow can be recorded with an optical probe or with a whole-field method. The differences between these two principles, their respective advantages and shortcomings have been discussed in Section 1.2. For determining the velocity of tracer particles, optical probes either make use of the Doppler effect in the scattered radiation (laser–Doppler anemometry), or they measure the time-of-flight of particles between two narrow light beams (e.g., laser–dual-focus method). It is interesting to note that the effect of laser–Doppler arrangement can be described also in terms of optical interferometry. The particle then passes through a system of maxima and minima of the optical radia-

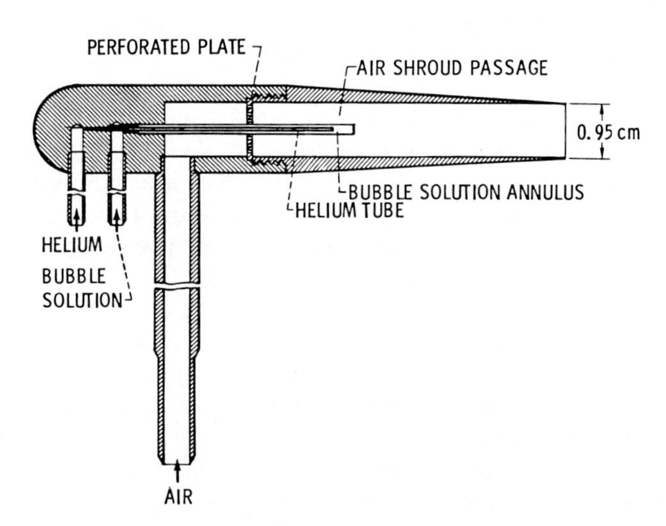

Fig. 2.19 Head of helium-bubble generator. (From Colladay and Russell, 1976. Reproduced with permission from The American Society of Mechanical Engineers.)

tion, and this description makes the method equivalent to a number of the optical tracking systems in which the image of a tracer particle is made to pass over an optical grating, thereby producing an oscillating light signal whose frequency is a measure of the particle's velocity (Gaster, 1964; Ushizaka and Asakura, 1983).

Our main interest is of course in the whole-field methods, which provide information on the distribution of the velocity at a particular instant of time in a whole field, or a whole plane in the flow. For reasons of signal intensity, tracer particles used for these methods usually are larger than those providing the signal for optical probes, and information on turbulent fluctuations is normally not available. The application of a whole-field method needs to arrange for an appropriate illumination of the field seeded with tracer particles, and for a recording system (i.e., a camera). Some of these technical needs have already been discussed in connection with dye and smoke visualization, and the systems used for seeding the flow with particles are also similar to the systems used in those cases. It will become evident that controlling the exposure time during the recording process is the most critical technical demand for measuring particle velocities.

2.2.3.1. Light Sheets

The most common illumination system is the light sheet with which the particles in a thin plane sheet in the flow can be illuminated (Véret, 1985). Such a sheet can be produced with a conventional light source (mercury lamp, projector lamp) and appropriate lenses and slit masks, or by expanding a laser beam in one plane by means of a cylindrical lens (Fig. 2.12). Continuous lasers with an output of 5 mW (He/Ne) up to several watts (argon) have been used for this purpose. The necessary light intensity depends on the amount of light scattered by the tracers and on the chosen exposure time. The application of discontinuous light sources (flash lamp, electric spark, ruby laser) allows one for controlling the exposure time by the generated light pulse. A light sheet usually is several centimeters high and about 1 mm thick.

With the light sheet one visualizes the two velocity components in one plane. Emrich (1983) calls the record of such a plane a "time window" within which the velocity can be determined simultaneously for many points. Shifting the sheet in a direction normal to its plane enables one to scan a three-dimensional flow, provided that the flow is stationary over the test period. However, the third velocity component, normal to the plane, is not recovered in this way. That would need to illuminate light sheets normal to the first set of sheets. An alternate solution has been described by Van Meel and Vermij (1961). Instead of one sheet, they use several parallel sheets with different colors. The motion of the tracer particle in the third dimension is indicated by a change of its apparent color, when it passes through the various colored light ribbons. When the recorded pattern is evaluated, the question arises whether the magnification factor is constant over the whole observed plane, or in other words, whether the image of the plane is free from distortion. As a solution to this problem, Owen and Campbell (1980) have devised an optical grid, which can be superimposed to the original light sheet. A "laser manifold" separates a single helium–neon laser beam into several uniform parallel beams, which are made to coincide with the illuminated plane sheet. By orienting two manifolds mutually perpendicular in that plane, one generates the optical grid as a reference frame.

The tracer particles in the light sheet are observed or photographed in a direction approximately normal to the illuminated plane. If one takes a time exposure, either by using a mechanical shutter or a pulsed light source, each particle appears in the image plane as a streak whose length is proportional to the average local velocity (Fig. 2.18). This simple way of measuring a velocity is accompanied by a number of technical shortcomings. The most serious error source is the precision and control of the exposure time. The higher the flow velocity, the shorter is the needed exposure time. Mechanical shutters operate down to about 1 msec; with a chopper wheel one may get an exposure time of about one order of magnitude less; electrical sparks can be fired over a length of only 1 μsec; and the pulse of a ruby laser can have a length of about 10 nsec. A compromise has to be found if the velocity in the observed plane varies over a great range. The streaks made by the fast tracer particles expose the film in a certain point at a shorter time, and the photographic contrast of these streaks, therefore, is lower than that of the short streaks generated by slow tracers. In a field with reversed flow the flow direction cannot be recognized without any additional means. In the beginning of a time exposure, Mezaris *et al.* (1982) therefore activate an additional flash, which creates a bright dot at the beginning of each particle streak.

The measurement of the length of the streaks or the particle displacement during the time exposure is most often performed from the negative. The accuracy of the so determined velocity is governed by the accuracy of the length measurement (e.g., 0.01 mm) and the precision of the exposure time. It is very difficult to give a general picture of the errors associated with these velocity measurements. Image processing provides new and additional means for the data evaluation. Dimotakis *et al.* (1981) have scanned and digitized the particle streak pattern from the photographic negative so that this pattern can be reproduced as a digital image. One can then apply special computer programs to this data field, for example, superimpose a constant velocity so that the streak pattern becomes visible in a moving frame, or derive stream function, vorticity, or other parameters by an appropriate numerical treatment (Imaichi and Ohmi, 1983).

More accurate than a single time exposure and the resulting streak is a stroboscopic illumination generating two or more extremely short light pulses with a well-defined delay time between pulses. Each tracer particle is then represented by a series of two or more well-focused images. The velocity is obtained from the distances of these image points. Light sources appropriate for this purpose are oscillatory electric sparks or a ruby laser. Examples have been reported in which the number of pulses was three (Emrich, 1983), four (Weinert *et al.*, 1980), five (Denk and Stern, 1979), and even fifteen (Breslin and Emrich, 1967; see Fig. 2.20).

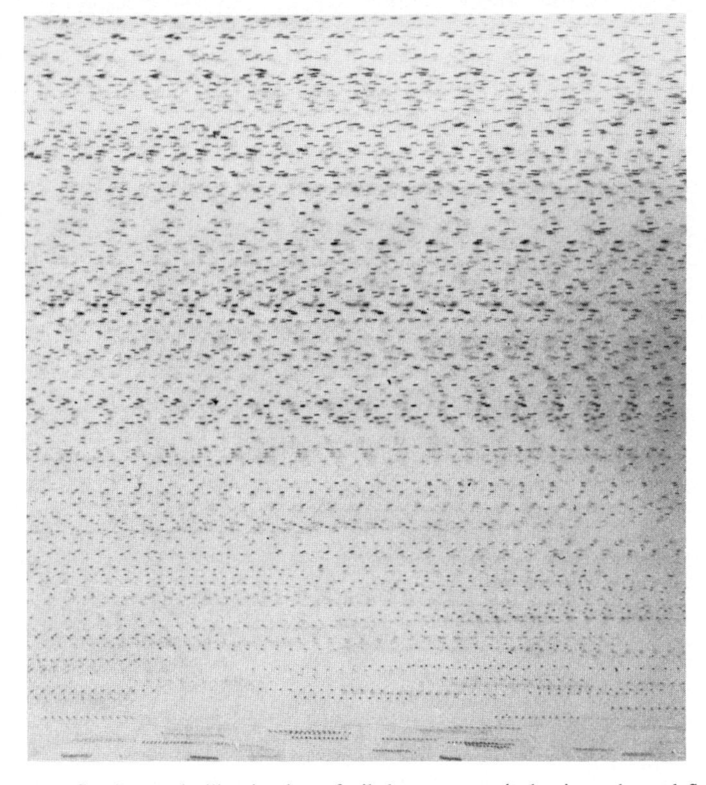

Fig. 2.20 Stroboscopic illumination of oil drop tracers in laminar channel flow. The area shown includes reflections in the glass wall and it covers approximately one-quarter of the region from the wall to the center. (From Breslin and Emrich, 1967.)

This stroboscopic illumination also enables one to identify those particles which leave the plane of illumination during the recording time: such particles will appear on the photograph with fewer dots than the due number of light pulses. Denk and Stern (1979) have devised a means for recognizing the direction of the tracer's velocity. Four equally long pulses are followed by a fifth short light pulse, which marks the end of the pulse series.

2.2.3.2. Stereoscopy and Holography

All three velocity components can be recovered by applying stereo-photography for the recording. Extensive literature is available on three-dimensional photography (e.g., McKay, 1951), and successful application to tracer particle fields have been reported (Elkins *et al.*, 1977; Kent and Eaton, 1982). A possible optical setup is shown in Fig. 2.21. Four front surface mirrors serve to produce on each film frame two simultaneous stereo images of the object field, which is seen under a relatively wide angle. Kent and Eaton use a high-speed cine camera. It is necessary to determine the three-dimensional position of a tracer particle in two successive film frames, and the time interval between the two frames must be

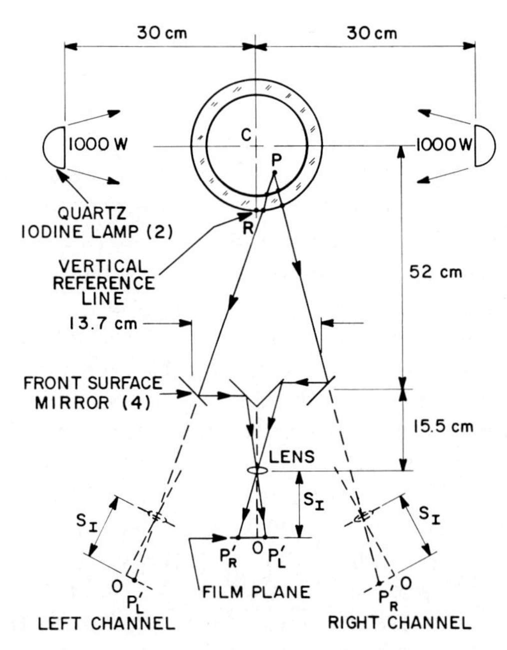

Fig. 2.21 Optical system for wide-angle stereo photography of flow tracer particles within a transparent cylinder engine model. (From Kent and Eaton, 1982. Copyright, Optical Society of America.)

known for measuring the velocity. The image of a particle, which must be identified in the two frames, appears with different coordinates in the two halves of each frame. Once these stereo image coordinates have been measured, a ray tracing procedure for each of the two channels delivers the exact spatial position of the particle. An error analysis shows that the error in the depth coordinate, that is, in the direction of the optical axis, is about an order of magnitude larger than in the two other coordinates. The evaluation of a whole field with the identification of each individual particle by visual observation is of course very time consuming. As in many other applications, this evaluation can be computerized by digital image processing (Chang *et al.*, 1985). The most critical part in such a procedure is the particle identification, which is more difficult for the computer than for the human eye. In any case it is required to have the particle seeding density low enough to facilitate the identification.

If the time for observation is long enough, the third dimension in the position of a particle can also be measured by means of observation through a microscopic lens system with short focal length (Berezin and Kudrinskii, 1977; Emrich, 1983). Such a system had already been used by Fage and Townend (1932) for tracing the flow close to a solid wall. In this case, that is, if the object of the study is the wall boundary layer, the wall itself can serve for providing the three-dimensional information. If the wall bounding the flow is made a surface mirror, each particle generates two images on the film, and a geometrical ray analysis can deliver the three spatial coordinates (Orlov and Mikhailova, 1978).

Holography is a means to freeze a three-dimensional picture of the flow scene. Upon reconstruction the three-dimensional record is available for detailed analysis of the spatial position of individual tracer particles. If two successive holograms are taken with a double-pulsed ruby laser, the vector change in tracer position can be determined and divided by the time interval between pulses to yield the vector velocity. Possible holographic arrangements have been described by Trolinger (1974). The interference pattern on the hologram produced by one single spherical particle consists of several concentric circular fringes; these are nothing but the particle's diffractive image caused by the interference of the scattered light with the undisturbed reference light. This diffraction pattern in the holographic plane allows for a direct determination of the particle's spatial position, since the radii of the interference fringes are unique functions of both the particle's size (which has to be known) and the particle's distance from the recording plane. This provides a means to measure three-dimensional velocity fields from a double-exposed hologram without the usual reconstruction (Menzel and Shofner, 1970).

Velocity measurements from a holographically reconstructed image are, of course, more easily related to the experience than the observation of the diffraction pattern. A holographic setup for double-pulse recording is given in Fig. 2.22. The reconstructed real image, formed on a ground glass, shows the true flow scene, only with a double image of each tracer particle. The scene can be viewed and scanned in depth for evaluation with a microscope lens. The exactness of the velocity measurement depends on the accuracy with which the individual particles can be identified and localized in the reconstruction. Several parameters, such as particle size, particle concentration (seeding rate), and dimensions of the optical system are contributing to this point. The resolution in focal depth is limited by diffraction in the holographic procedure as well as by the quality of the zooming lens, and it can best be achieved to an accuracy of about 10 particle diameters.

Three-dimensional velocity measurements of tracer particles and in two-phase flows have been performed with such arrangements [e.g., by Trolinger *et al.* (1969), Lee *et al.* (1974), Charwat (1977b), and Ewan (1979)]. Of special interest is the simultaneous measurement of the particle size from the hologram (Fig. 2.23).

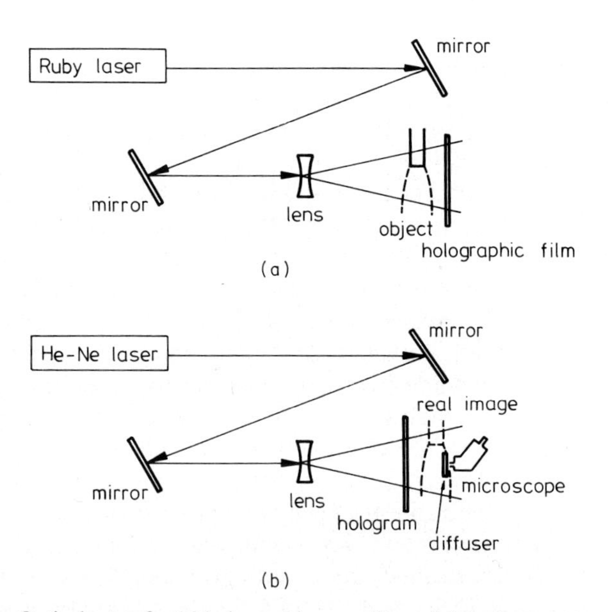

Fig. 2.22 Optical setup for (a) holographic recording and (b) subsequent reconstruction of a flow seeded with particles. In the reconstruction, the real image of the tracer particles is observed with a microscope. (See Lee *et al.*, 1974.)

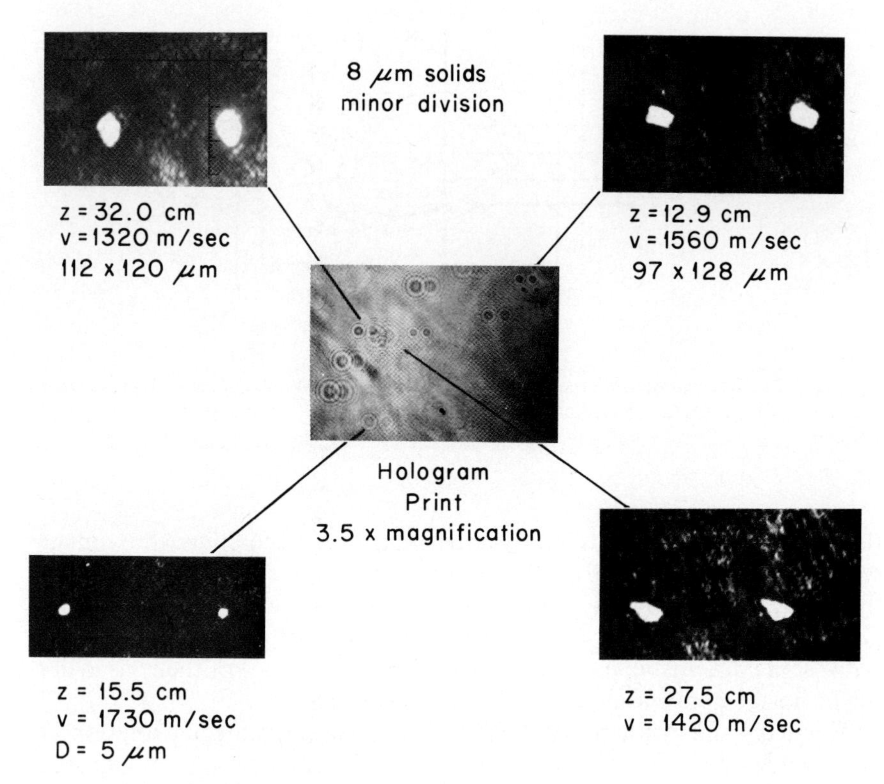

Fig. 2.23 Double-exposure hologram of a high-speed dust flow and four reconstructions obtained by focusing the camera at different planes; z = constant in the flow field. Particle size and velocity v are indicated for each reconstruction. Double-pulse separation is 0.48 μsec. (Courtesy of J. D. Trolinger, ARO Inc., Arnold AF Station, Tennessee.)

2.2.3.3. *Transparent and Nontransparent Test Section Walls*

If the flow is observed through a transparent viewing window it might be necessary to account for the refraction of light when the position of the tracer particle is determined. The respective correction is easy for a plane window, but the problem is more difficult to solve for curved optical boundaries. The problem is also present in laser–Doppler anemometry; Bicen (1982) has described ways for correction that are different for the axial, tangential and radial velocity component.

In the case of a liquid test fluid, an alternate solution to this problem may totally eliminate the refraction effects. What one needs is a liquid, which has the same refractive index as the transparent bounding walls (e.g., Plexiglas or Perspex). A flow channel of a complicated geometry is

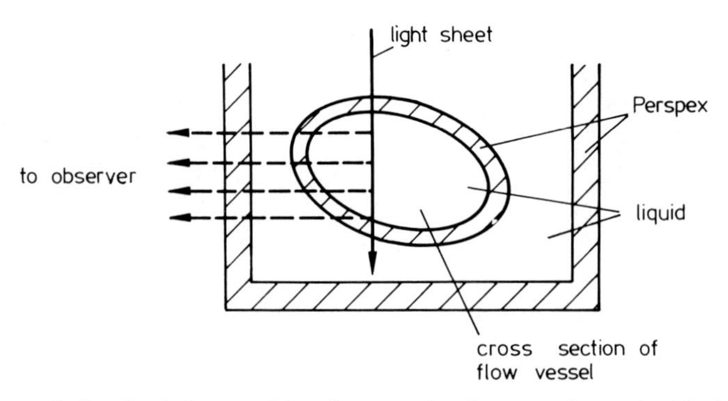

Fig. 2.24 Refractive index matching: flow vessel and surrounding tank with plane walls are filled with a liquid whose refractive index matches that of the wall material, e.g., Perspex.

then imbedded in a tank filled with the same liquid and bounded by plane walls (Fig. 2.24). Donnelly (1981) has listed a number of liquids, most of them organic compounds, whose refractive index is close to that of Plexiglas, $n = 1.5$ (see Table 2.4). These liquids are not all pleasant to work with, some are unstable or even unsafe, and the values of their viscosity might not be appropriate for a specific experiment.

Hendriks and Aviram (1982) have found that an aqueous solution of zinc iodide (ZnI_2) is a liquid whose index of refraction can be adjusted over a wide range of values overlapping with those of machinable and

TABLE 2.4

Liquids to Match the Refractive Index of Perspex (Plexiglas, Lucite) (n_D = 1.50–1.52)[a]

		Refractive index, n_D	Density, ρ (kg/m^3)	Viscosity, μ (Centipoise)
Isophorone	$C_9H_{14}O$	1.4881	$0.923 \cdot 10^3$	2.4
Dipentene	$C_{10}H_{16}$	1.480	$0.842 \cdot 10^3$	1.7
p–Xylene	C_8H_{10}	1.495	$0.861 \cdot 10^3$	0.65
O–Xylene	C_8H_{10}	1.505	$0.88 \cdot 10^3$	0.8
Dimethylphtalate	$C_{10}H_{10}O_4$	1.514	$1.189 \cdot 10^3$	
Monochlorobenzene	C_6H_5Cl	1.525	$1.106 \cdot 10^3$	0.8

[a] From Donnelly (1981).

castable transparent solids, and whose viscosity is also adjustable and low enough to generate relatively high Reynolds numbers in a model experiment. Refractive index and viscosity of this solution are related to each other (Fig. 2.25) and depend on the concentration of ZnI_2 in water. Unfavorable properties of this liquid are the high price of ZnI_2 and its hygroscopic nature. Hendriks and Aviram also discovered that the liquid is electrochromic; that is, a dark-brown trace of iodine is produced at the anode, which allows for an additional flow visualization in form of colored streaks, if an electric control system like described in Section 2.3 is applied.

X-ray photography is a means for visualizing the flow in a facility the walls of which are not transparent for the visual radiation. Hornemann (1982) has thus photographed the propellant gas flow inside a gun barrel, and by applying a double-pulsed x-ray flash he measured the velocity of powder particles in the barrel. X-ray photography is of course a technique quite familiar in medicine, and although velocities have barely been measured in this way, it has been used to visualize flow phenomena in blood vessels (Fig. 2.26).

2.2.4. *Velocity Measurement by Speckle Photography*

Laser speckle photography is a method that has extensively been used for measuring displacements and surface deformations in solid-state mechanics. The optical phenomenon of laser speckle has been explained, e.g., by Françon (1979). In recent years, use has been made of this phenomenon for measuring flow velocities in fluids, which have been

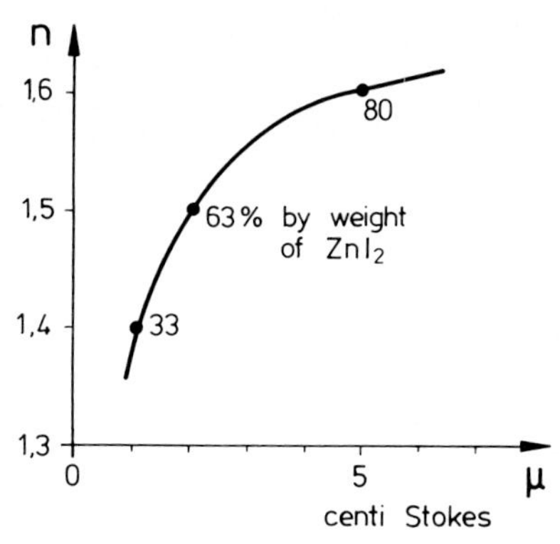

Fig. 2.25 Refractive index n and viscosity μ of aqueous solutions of zinc iodide. (From Hendriks and Aviram, 1982.)

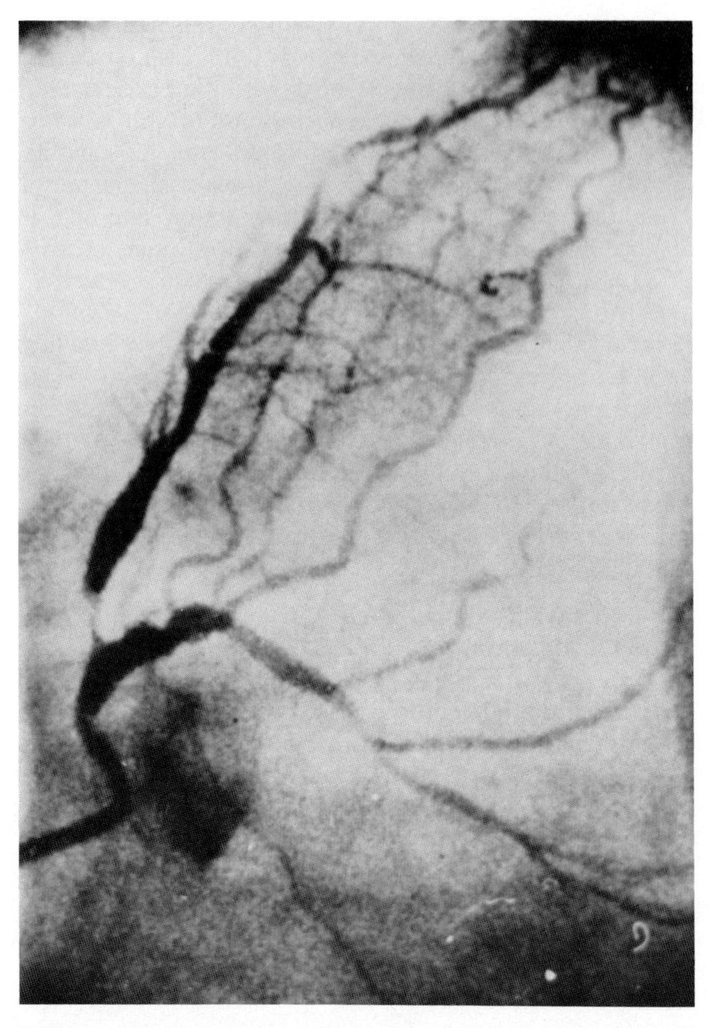

Fig. 2.26 A x-ray photograph showing a portion of the cardiovascular vessel system. A stenosis (a severe constriction) is seen in the lower left-hand corner. (From G. Blümchen, Klinik Roderbirken, Opladen, Germany.)

seeded with tracer particles (Barker and Fourney, 1977; Dudderar and Simpkins, 1977; Grousson and Mallick, 1977; Lallement *et al.*, 1977). The technique is a special recording method rather than a direct visualization method. We discuss it here separately and in detail, because it allows for both quantitatively measuring the velocity, and generating a visual pattern of a whole field once the record has been taken.

For explaining the principle we consider a bright object point, which is imaged onto a recording plane (Fig. 2.27). Suppose that the object point is displaced by a small distance ds; then, the image point is displaced in the same direction and by the same amount of ds, if we assume the imaging ratio to be 1 : 1. In a photographic double exposure, one exposure taken

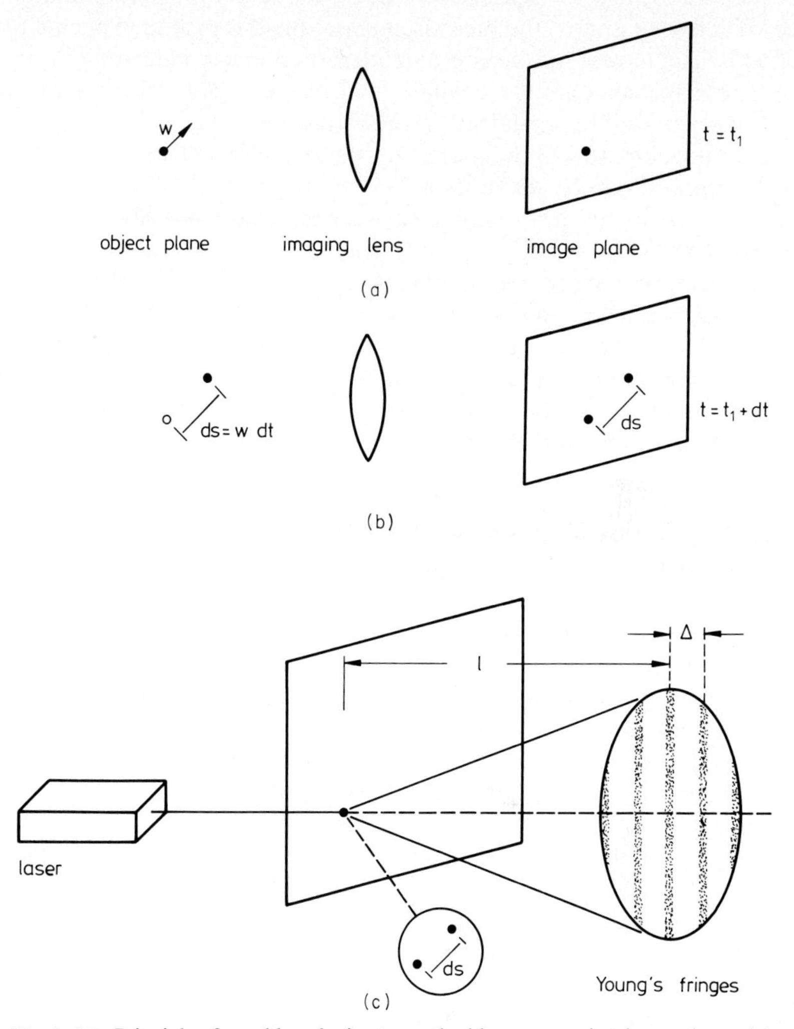

Fig. 2.27 Principle of speckle velocimetry: a double exposure is taken at time t_1 (a) and $t_1 + dt$ (b) of the object point moving at a velocity w. For the reconstruction (c) the laser beam illuminates a small circular area including the two images of the object point. A system of interference fringes is formed in a plane at distance l from the illuminated negative of the double exposure.

before and the second after the displacement, the object point appears twice, separated by ds as shown in Fig. 2.27. In our application to a fluid flow, the object point represents a single tracer particle moving along the flow with velocity w. The displacement of the particle between the two photographic exposures is $ds = w\,dt$, where dt is the time interval separating the two exposures. With known time interval dt, the determination of the velocity w requires the measurement of the distance ds. Speckle photography is a means for measuring both direction and magnitude of small in-plane displacements for a whole field of view. The principle of such measurement will be explained in the following.

After development of the double exposure the two object points appear on the plate as two bright spots on a dark background. These two points are now illuminated from behind with a thin laser beam (Fig. 2.27). For the propagation of the light in the space to the right of the plate, the two illuminated spots act as two point sources; they emit light into two cones. In the overlapping regime of the two cones the light from the two sources can interfere. Parallel and equally spaced interference fringes ("Young's fringes") can be observed in a plane normal to the optical axis and at a distance l from the plate with the two sources. The fringes are normal to the direction of ds and separated by the spacing Δ:

$$\Delta = l\lambda/ds,$$

where λ is the laser wave length. The measurement of ds is thereby reduced to the measurement of the fringe spacing Δ in a system of Young's interference fringes.

Although the situation with one single particle is far from being realistic, it can be used for explaining a certain limitation of the method. For this purpose we have to take into account that the particle as the object point and its image have a particular diameter, say d_p. Then, the lower limit for the displacement to be measured in this way is given by the requirement $ds \geqq d_p$. There exists also an upper limit when the fringe spacing Δ becomes too small, that is, when the fringe system is too narrow. A rule of thumb is that this upper limit is reached for $ds \cong 20d_p$.

Under normal circumstances the illuminated object plane contains a great number of tracer particles, which move between the two exposures. The laser beam directed onto the developed double exposure has a diameter of about 1 mm or less. In the circular area illuminated by this beam we may have several pairs of image points. If all these points have been displaced by the same amount and in the same direction, the resulting system of Young's fringes is the same as if only one pair had been illuminated. Figure 2.28 is a schematic representation of such a situation. Three groups of tracer particles are moving in the object plane at different veloc-

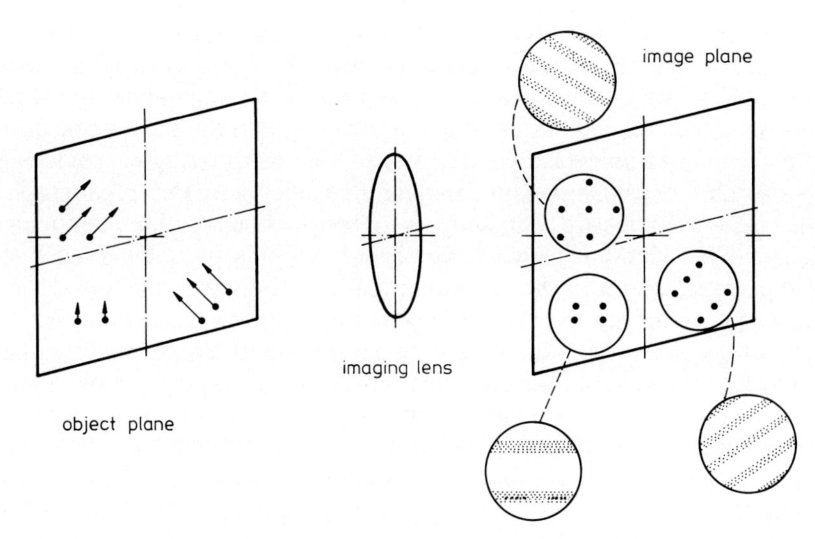

Fig. 2.28 Schematic representation of speckle velocimetry with three groups of object points moving at different velocities. The different patterns of Young's fringes obtained in the reconstruction process are indicated for each group.

ities. The laser beam used for the evaluation illuminates the images of each group in the developed double exposure. Also shown are the respective fringe systems; fringes are normal to the velocity direction, and their spacing is the narrower the higher the particle speed.

If the pairs of image points illuminated simultaneously by the laser beam have not all experienced the same displacement, then different systems of Young's fringes will be generated and they overlap each other. The observable pattern might be inappropriate for evaluation, the fringe pattern is "decorrelated". The assumed fringe pattern in Fig. 2.28, which gives a maximum of contrast between fringes, is well correlated. One may conclude that a small deviation in amount and direction of the displacement of the simultaneously illuminated image points causes a certain loss of fringe contrast and correlation. If we tolerate this degree of decorrelation and measure the fringe spacing and direction in this system of reduced contrast, we attribute a mean value of the displacement to all object points whose double image is illuminated and located in the (circular) cross section of the thin laser beam. A reduction of this cross section, for example, by employing a focusing lens, is desirable for improving the local resolution.

For further discussion we consider a real flow experiment. A plane in the flow field is illuminated by means of a thin light sheet (see Section 2.2.3.1). The illuminated tracer particles moving in this plane are the "object points." Most suitable for taking a double exposure of the illuminated sheet is the use of a double-pulsed ruby laser as a light source. Then, the width of the single pulses as well as the time interval between pulses can be controlled short enough to allow for a tolerable value of the displacement to be measured. The use of a laser source now provides an opportunity for explaining the origin of the name "speckle" photography. The tracer particles are, as mentioned in the previous sections, scatterers; that is, they reflect the incident laser light in various directions. The image of the object plane in the recording plane is therefore not formed in a simple imaging process. Due to the coherence of the scattered radiation, the pattern observed in the recording plane is the result of multiple interference between light rays arriving there from various directions. Such a pattern is called a speckle pattern; the individual "speckles" of this granular structure correspond to what we have described as the "image points" of the tracer particles. The diameter of an individual speckle, which is d_p can be determined from diffraction theory (Françon, 1979).

The double-exposed photographic plate contains information on the distribution of the (two-dimensional) velocity vector in the illuminated plane. For evaluation the developed plate is scanned point by point with a laser beam like that indicated in Fig. 2.27; this procedure is called the reconstruction process: the information is reconstructed from the "specklegram" by means of an appropriate evaluation method, here the "point-by-point analysis." For each illuminated point one obtains a pattern of Young's fringes. From the measured fringe spacing and fringe direction one determines the two velocity components in the respective point of the illuminated plane. Such a system of Young's fringes is shown in Fig. 2.29. The velocity (or displacement) vector is normal to the fringe direction. Due to the intensity distribution in the cross section of a laser beam, the intensity decreases in the radial direction. The dark area in the center of the picture is caused by a diaphragm blocking off the direct radiation of the laser beam.

From the fringe pattern it is not possible to determine whether the particle is moving in the positive or the negative direction. By means of an additional displacement between the two exposures one could overcome this difficulty. This would be in analogy to the frequency shift in laser–Doppler anemometry.

The velocity component normal to the illuminated plane is not recovered. If a particle moving in this third direction is in the plane at the instant of one of the two exposures, it will appear as a single image in the double exposure. The light scattered from such particles contributes to the ground noise reducing the visibility of the fringes. From the preceding discussion on signal decorrelation due to small deviations in the displace-

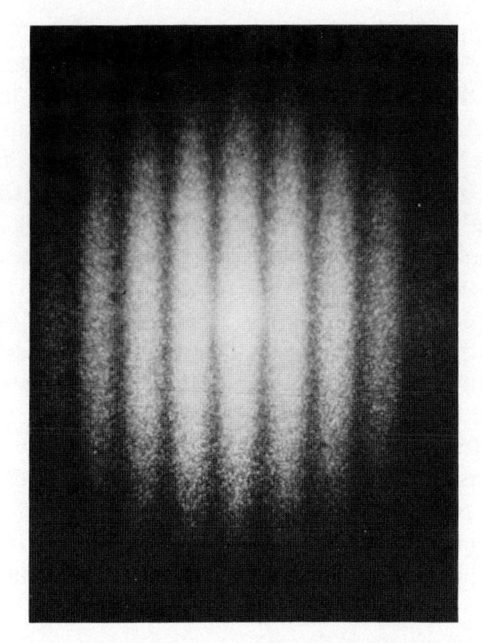

Fig. 2.29 Photographed system of Young's fringes.

ment vector it is evident that noise (in the sense of a loss of fringe visibility) is also generated by turbulent fluctuations in the velocity vector, as well as by a nonuniform distribution in the particle size. At the same time, this noise is carrying information, and it has been speculated whether this "noise" could be decoded to obtain, additionally to the measurement of mean velocity, information on turbulence characteristics (Hinsch *et al.*, 1984) or on particle size and concentration (Genceli *et al.*, 1980).

The sharpness of the Young's fringes in the reconstruction procedure can be improved if these fringes are formed by multiple-beam interference rather than by two-beam interference. This has been verified by Iwata *et al* (1978) who took multiple exposures for recording the velocity field. Illumination for the recording can be provided even from a white source (Bernabeu *et al.*, 1982; Suzuki *et al.*, 1983); interference then is not involved in the recording, which is now a simple imaging process, and the question has been raised whether "white-light speckle photography" is related to any formation of speckles.

Since the speckle pattern to be recorded has a very fine structure, and since it is necessary to resolve displacements of image points of the order of microns, high-resolution holographic film material has been used in many applications. The information stored in the record ("specklegram") is a very dense distribution of data on displacement vectors or velocity, and scanning of the specklegram in the point-by-point analysis for reconstruction of the data values can be a very time-consuming procedure if it is done manually. The rational analysis of extended data fields, therefore, requires the use of an automated processing system. In these systems the pattern of Young's fringes is digitized, and fringe spacing and fringe direction are determined by a computer for each point of the scanning procedure (Meynart, 1982; Erbeck, 1985; see Fig. 2.30).

The speckle technique allows for an additional way of reconstruction resulting in a real visualization of the flow field. This analysis of the data is called "spatial filtering", and a schematic set-up for this reconstruction is shown in Fig. 2.31. The developed double exposure ("specklegram") now is illuminated with an expanded laser beam. The specklegram is imaged by means of a lens in a plane ("Fourier plane") where, in the aforementioned point-by-point analysis, the pattern of Young's fringes had been observed. Here, an opaque screen is placed in this plane having a transparent hole at an off-axis location. A small amount of the light can

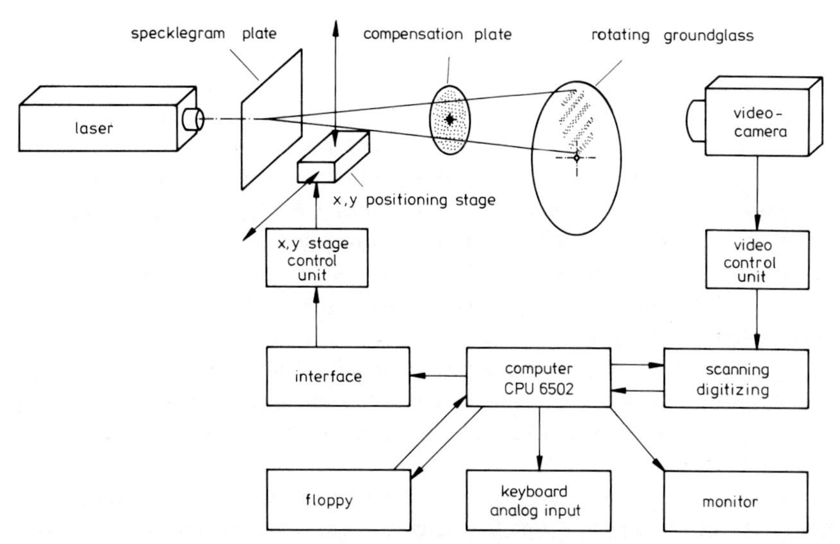

Fig. 2.30 System for automated scanning and evaluation of a specklegram. A system of Young's fringes is formed on a rotating groundglass where the laser noise is damped. A special filter plate compensates for the changes of mean intensity across the halo. The fringe system is viewed by a TV camera whose signal is digitized and evaluated by a computer. The computer also controls the x, y positioning stage on which the specklegram is mounted. (From Erbeck, 1985. Copyright, Optical Society of America.)

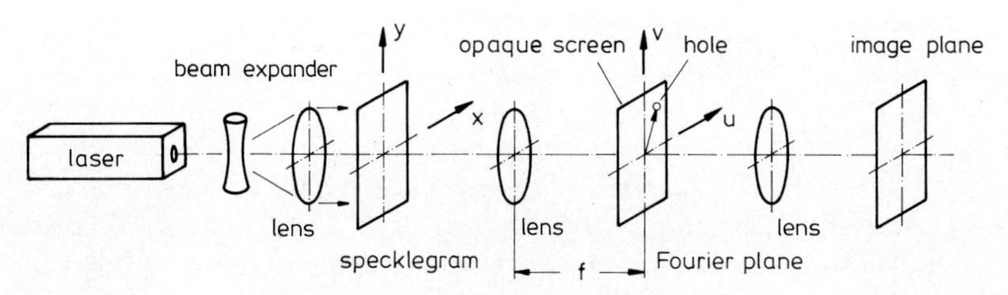

Fig. 2.31 Reconstruction by means of spatial filtering of the specklegram.

pass through the hole, and a second lens forms an image of the specklegram in the final image plane. Since every point of the specklegram is illuminated by the expanded laser light, the Fourier plane is covered by an infinite number of overlapping fringe systems. Light can pass through the hole if the hole, in the Fourier plane, is located on a bright interference fringe. It turns out that this requirement is met by such systems, which are normal to the radius combining the hole and the axis, and for which the ratio of the distance from the hole to the optical axis and the fringe spacing is an integer number. As a consequence, only such points appear illuminated in the (final) image plane for which the flow velocity is of certain value and direction in the respective object points of the flow plane. The points illuminated in the image plane lay on curves or fringes of equal velocity component. The velocity difference from fringe to fringe (not to confuse with the Young's fringes!) is constant. The amount and the direction of the velocity component visualized by means of these "equivelocity fringes" depends on the location of the hole in the Fourier plane.

Figure 2.32 shows a system of equivelocity fringes in the field of a plane internal gravity wave, which has been produced by a cylinder moving upward through density-stratified saltwater. The observable velocity fringes are at the same time curves of equal phase in the wave system. Locating the hole at a different position in the Fourier plane, results in visualizing a different velocity component. Equivelocity fringes have been visualized, e.g., by Meynart (1980) and by Suzuki *et al.* (1983). This reconstruction by spatial filtering suffers from the low light intensities that have to be managed.

Applications of speckle photography for velocity measurements are still not very frequent. An apparent limitation is the availability of an ultrashort light source for generating the double exposure. This problem is smaller the lower the flow velocity to be measured; this fact explains successful applications of the method to Bénard convection (Simpkins and Dudderar, 1978; Meynart, 1980) and a particular biological flow problem (Fercher and Briers, 1981). However, the general suitability of speckle

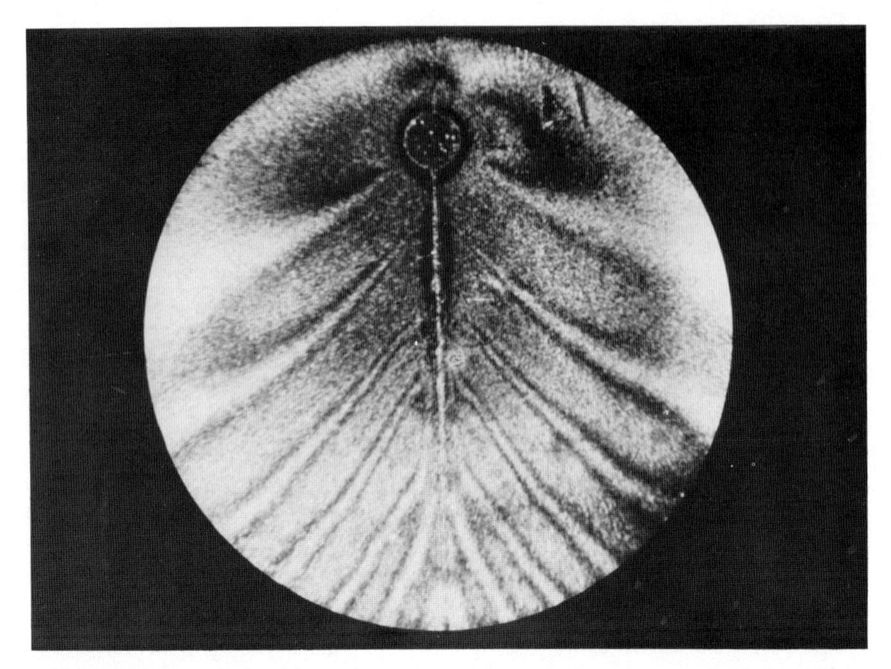

Fig. 2.32 Equivelocity fringes in the field of an internal gravity wave behind a cylinder moving upward through stratified saltwater. (From Gärtner *et al.*, 1986.)

velocimetry for gas flows with higher speeds has also been proven (Meynart, 1983).

2.3. Visualization of Velocity Profiles

It has been mentioned in Section 2.1.2 that methods exist for producing dye in certain liquids along a prescribed curve and with exact temporal control. Such methods will be discussed. If the dye is generated along a line perpendicular to the mean flow direction, it may enable one to visualize local velocity profiles by observing the distortion of such dyed line in the flow. In Section 2.1.3 we discussed such a method for air flows, namely the smoke wire. Later, in Section 4.2., we shall learn about another method of this type, which is also used for air: the spark discharge method, which does not rely on the presence of a wire. The same situation applies to water flows. A number of methods exist for which the "prescribed curve" is defined by a thin, straight wire, and the production of dyed markers has to do with an electrochemical reaction in the liquid; and it is also possible to generate the linetracer by means of a photochemical reaction (i.e., without a wire) (Section 2.3.4).

In the electrochemical process for which the wire serves as an electrode, the flow is marked either by material separating from the wire surface (Section 2.3.1), by electrolytically generated hydrogen bubbles (Section 2.3.2), or by a pH-indicator, which changes color due to the alteration of the pH-value in the immediate neighborhood of the wire electrode (Section 2.3.3). As in the case of the smoke wire, it is also possible to visualize streamlines (or filament lines) if the marked portions of the flow are observed in a direction parallel to the wire line.

2.3.1. Dissolution of Electrode Material

Wortmann (1953) probably was the first to report on the use of an electrochemical reaction for controlled flow visualization and the measurement of velocity profiles. His tellurium method has later been reviewed by Eichhorn (1961). The working fluid is water whose electrolytic conductivity can be increased, if necessary, by a respective additive. Two electrodes are introduced into the fluid; the cathode is a thin wire of tellurium and is placed in the flow regime under study. With an electric voltage applied between the electrodes, tellurium ions with a double negative charge are dissolved from the wire surface. With oxygen present in the fluid, the tellurium ions are brought into the state of colloidal suspension appearing in the fluid in form of black dye. The suspended particles have diameters of about 1 μm, a low settling rate (0.1 mm/sec), and a low rate of diffusion ($<10^{-2}$ mm/sec).

With a short electric pulse applied, the dissolved tellurium cloud is a thin cylinder adjacent to the wire. If the cathode wire is oriented normal to the mean flow direction, the edge of the tellurium cloud moves with the flow, and its position on a photograph represents the velocity profile of the flow, which has passed the wire (Fig. 2.33). Wortmann (1953) gives a number of detailed instructions for improving the contrast and the stability of the tellurium cloud. The colloidal state of the cloud is stabilized by increasing the pH value of the solution, for example, by adding a small amount of KOH until a value of pH = 9 or 10 is reached. The contrast is improved by the addition of a few drops of H_2O_2; one thereby makes available the additional O_2 molecules necessary for the formation of the colloidal suspension. The contrast of the cloud also depends on the magnitude and duration of the applied voltage, and on the conductivity of the electrolytic solution; these quantities determine the number of tellurium

Fig. 2.33 Velocity profile in the boundary layer over a flat plate as visualized with the tellurium method. The tellurium wire is seen on the left-hand side of the figure. (From Wortmann, 1953.)

ions produced by the electric pulse. Typical values are a dc voltage of 300 V at a current of about 1 A/cm of cathode wire length. The cathode wires are either drawn from metallic tellurium; or tellurium is evaporized and coated onto a thin steel wire. The wire has a limited lifetime; it may last for about 100 experiments.

The wake of the cathode wire causes a deficiency in the velocity measurement. This error source is associated with all methods employing electrode wires, and it will be discussed in more detail for the hydrogen-bubble technique. Another shortcoming of the method is that the production of tellurium is an irreversible process, contaminating the test fluid more and more with time. The technique has found only a limited number of applications for studying boundary layer and convective flows (Brooke-Benjamin and Banard, 1964; Schetz and Eichhorn, 1964; Wortmann, 1969), but it has played a pioneering role among the electrochemical visualization methods.

Dissolved electrode material is also the flow marking material in the "electrolytic precipitation method" developed by Honji (1975) and again described by Taneda *et al.* (1979). The test body itself whose surface is coated with a thin layer of solder or tin is used as the anode in an electrolytic circuit. The cathode can be placed at any desired position in the flow field. Normal tap water is an appropriate working fluid. Upon applying a dc voltage of about 30 V, a cloud of white dye separates from the anode surface and moves with the flow. It marks the trace of the separating boundary layer, as demonstrated for different cylindrical configurations (Kiya *et al.,* 1980; Honji, 1981). The dye is a metallic salt formed in the electrolytic process; it consists of nearly spherical particles of low sedimentation velocity. When the water velocity exceeds about 5 cm/sec, it becomes difficult to produce a sufficient amount of dye; that is, the visible contrast of the dye decreases with increasing fluid velocity, and the technique is appropriate for slow water flows only.

2.3.2. *The Hydrogen-Bubble Technique*

It has been reported that the hydrogen-bubble technique has been found incidentally when researchers in a laboratory played with the tellurium method. The basic principle of the technique is the electrolysis of water. If a dc voltage is applied between two electrodes in water, hydrogen bubbles are formed at the cathode and oxygen bubbles at the anode. Since the hydrogen bubbles develop with much smaller size than the oxygen bubbles, only the hydrogen bubbles are used as tracers. If the cathode is a long, fine wire placed normal to the mean flow, and if a short electric pulse is generated, a row or column of bubbles is formed along the wire (comparable to the tellurium cloud), carried away with the flow, and deformed according to the local velocity profile. By pulsing the voltage at a constant frequency, one produces several successive columns of bubbles, which mark in the flow curves separated by a constant flow time. These curves are called "time lines," and it is in this mode that the hydrogen-bubble technique is most frequently employed.

The earliest reference on the hydrogen-bubble technique apparently is that of Geller (1955). The description of Clutter and Smith (1961) contributed very much to the growing interest in the method. A quantitative analysis has been developed by Schraub *et al.* (1965), who also recognized that this technique is an appropriate means for studying boundary layer flows at low water velocities. Davis and Fox (1967) revised the technique and applied it to the measurement of velocity profiles in circular tubes. The easy control of generating the flow tracers makes the hydrogen-bubble technique appropriate for lecture demonstrations (Tory and Haywood, 1968). Today, the apparatus is commercially available.

In a conventional arrangement, a thin, fine wire placed in the flow is used as the cathode (negative electrode) of a dc circuit for electrolyzing the fluid, in most cases water. Wires are made of platinum or stainless steel with diameters of the order of 0.01 to 0.02 mm. The anode with arbitrary shape is placed at some other location in the flow. Normal tap water may serve as the electrolytic fluid; if the water is too soft, one may add sodium sulfate in order to increase the electrolytic conductivity. Hydrogen bubbles are produced at the cathode wire if a dc voltage is applied between the electrodes. The bubbles are carried away with the flow. Owing to the buoyancy the bubbles necessarily rise in the flow during their motion with the fluid, and it is desirable to have their rise rate as low as possible. If the bubbles can be generated very small in size, the Reynolds number associated with their rising motion is low, and the rise velocity is determined by Stokes' law [see Eq. (2.6)]. Hence, it is in any case desirable to generate the bubbles as small as possible. A rule of thumb is that the bubble size is of the order of the diameter of the generating cathode wire. Matsui *et al.* (1979) who observed with a microscope the process of bubble formation found, that this rule applies to conventional systems when smaller bubbles had coagulated immediately after their formation.

However, the bubble size also depends on other parameters, for example, on the conductivity of the fluid and on the applied voltage or electric current. With tungsten wire, 5 to 10 μm in diameter, electric pulse widths of 1 to 4 μsec, and voltages of 100 to 600 V, Matsui *et al.* (1979) produced bubbles that apparently did not coagulate and were much smaller in size than the wire diameter. The electric currents must be large enough to obtain sufficient optical density of the bubbles. Care should be taken to obtain a uniform voltage and thereby a uniform bubble production along the wire. An appropriate electric circuit is needed to provide uniform bubble production during the experiment and to avoid the formation of large bubbles in the initial part of each experiment. It has been observed that the quality of the bubbles changes after a few minutes of continuous operation, presumably as the result of the deposition of contaminating materials on the cathode. This may be corrected by reversing the polarity in the electric circuit for about 30 sec.

If the cathode wire is placed normal to the mean flow direction, and if a short electric pulse is generated in the circuit, a row of bubbles is produced along the wire. They mark a line of fluid elements whose position coincided at a given instant with the position of the wire. Any later location of these rows of tracer particles is called a "time line," which is a measure of the local velocity profile. By pulsing the voltage with a known frequency one obtains several consecutive rows of bubbles (Fig. 2.34). The local velocity can be determined by measuring the distance between the bubble rows and dividing by the time between the electric pulses.

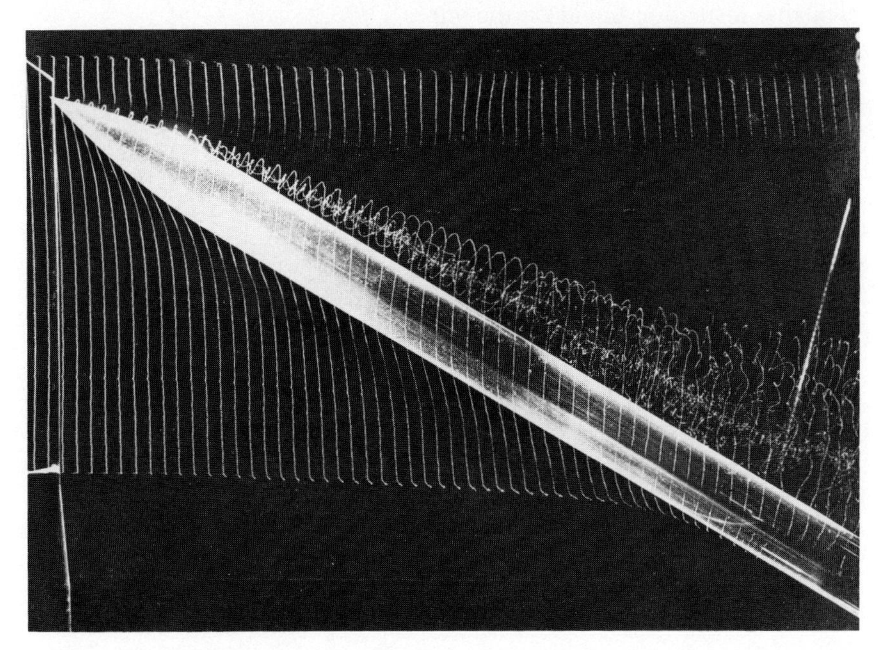

Fig. 2.34 Hydrogen-bubble time lines visualizing the flow over an ogive-nosed circular cylinder at high angle of attack. (Courtesy H. Bippes, M. Hartmann, DFVLR Göttingen, Germany.)

Essential to the accuracy of this method is an electric power source, which yields precisely timed voltage pulses at a constant repetition rate. Clutter and Smith (1961) already demonstrated that ordinary 50- or 60-Hz ac may be used as the voltage source, thus producing alternating rows of hydrogen and oxygen bubbles at the wire, which is now by turns the cathode and the anode.

Such a system of time lines does not provide information about the particle trajectories because it is not known which tracer particles or points of the two subsequent time lines correspond to one another. Schraub *et al.* (1965) therefore developed a combined time-streak technique by using a cathode wire having short sections coated with insulation. If the applied voltage is pulsed at regular intervals, combined time-streak markers are produced (e.g., in form of initially regular squares of hydrogen bubbles) (Figs. 2.35 and 2.36). This combined time-streak technique can yield a more complete picture of complex flow fields, e.g., of the eddy or vortex structure in a nonlaminar flow field, which is a frequent test object of the hydrogen-bubble method. In order to make a partially isolated cathode wire, one masks those sections of the wire, which are not intended to remain isolated, and one may then spray a commercially available insulation liquid on the wire.

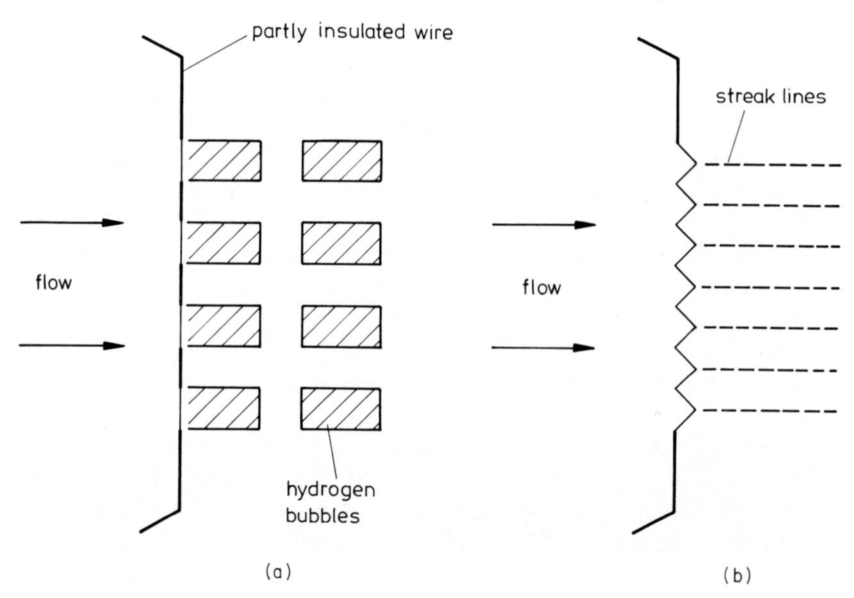

Fig. 2.35 Generation of (a) combined time-streak pattern with partly insulated wire and (b) streak lines with kinked wire.

Fig. 2.36 Time-streak marker technique applied to the flow in a contraction. Flow direction is from right to left. (From Schraub *et al.*, 1965.)

A kinked cathode wire allows for generating a streak (or filament) line pattern, because the bubbles will only develop from the kinks (Fig. 2.35). With two cross-lined wires (e.g., oriented at ±45° to the mean flow direction) one may produce a grid structure in which both time lines and streaklines can be identified by respectively connecting the nodal points of the grid pattern (Matsui *et al.*, 1979).

The usual illumination and recording systems are similar to the arrangements described earlier for the observation of tracer particles or dye lines. Several investigators (e.g., Offen and Kline, 1974; Bippes, 1977) found an angle of 65° between illumination and viewing direction most suitable. This might have to do with the scattering characteristics of the hydrogen bubbles. Conventional photography or cinematography is used for recording; a video camera is of advantage for the further processing of the flow pictures (Smith and Paxson, 1983). Systems have been described in which the bubble velocity is determined without photography, either with some kind of laser light fence (Hearney and Davis, 1974), or by observing two bubble rows generated by two separated wires with controlled delay time (Iritani *et al.*, 1983).

The period of time during which the bubbles can be observed in the flow is limited by the dissolution of the hydrogen bubbles in the fluid. Diffusion of the bubbles increases with Reynolds number and is very rapid in turbulent flows. The application is therefore limited to lowspeed flows, maximum velocities being of the order of 20 to 30 cm/sec. Besides water, water–glycerine mixtures (Criminale and Nowell, 1965; Roos and Willmarth, 1969) and an aqueous sugar solution (Bhaga and Weber, 1981) have been used as the working fluid; this is of interest if the aim is to keep the Reynolds number low.

An elaborate investigation of the error sources and uncertainties associated to the hydrogen-bubble technique has been performed by Schraub *et al.* (1965). Velocity values are obtained by dividing the measured displacement of the bubble rows through a known time interval. This time interval is determined by the bubble-wire pulsing rate or by the camera framing speed to a high degree of accuracy. The severest errors are therefore contained in the measured displacement Δx of the bubbles. Several factors contribute to the total uncertainty in Δx. The magnitude of these particular contributions depends on the flow under study. It turns out that some operations, which are intended to reduce one of these error sources simultaneously increase the contributions from another imperfection. The individual error sources will be listed in brief.

The evaluation of a photograph made of the bubble motion is associated with a measurement uncertainty in the displacement Δx. Uncertainties of this type can be reduced by utilizing a longer time interval Δt between exposures or electric pulses. On the other hand, since the velocity is averaged over the time interval Δt, the uncertainty due to this averaging increases with Δt. A normal flow photograph does not show a displacement of bubbles out of the plane formed by the cathode wire and the mean flow direction. A movement of the bubbles out of this plane could be caused by buoyancy forces in the flow. A vertical rise velocity is superposed to the bubble motion in the flow. If one assumes the rise rate to be determined by Stokes' law, the rise velocity decreases with the diameter of the bubbles. If the flow field contains steep velocity gradients in the direction of bubble rise, the associated errors can be considerable, since the bubble is now entrained in a direction of changing flow velocity, and its motion deviates strongly from the movement of a neutrally buoyant fluid particle. In some cases, these errors can be avoided by a change of the flow geometry.

Although the overall flow pattern is not seriously disturbed by the presence of the cathode wire, the local disturbance, which is caused by the wake of the wire, and which affects the motion of the bubbles, can be great. The bubbles moving in the wake of the generating wire indicate the corresponding velocity defect. Due to the small diameter, the wire Reynolds number usually is low. From laminar wake theory one may conclude that the bubble velocity reaches the free stream velocity in a distance of about 70 to 100 wire diameters downstream of the wire. This means that the test regime in the flow should be at least 100 diameters behind the wire. If the wake flow does not remain laminar, the generated turbulence will disturb the bubble movement as discussed previously.

Probably the greatest achievements, in applying the hydrogen-bubble technique, have been made in studying the structure of turbulent boundary layers (e.g., Kline *et al.*, 1967; Kim *et al.*, 1971; Grass, 1971; Bippes and Görtler, 1972; Thomas and Rice, 1973; Offen and Kline, 1974; Smith and Metzler, 1983; see Fig. 2.37). Visualization of velocity fields has been reported for steady and unsteady pipe flow (Johnston *et al.*, 1972; Clamen and Minton, 1977; Kato *et al.*, 1982), jets and their interaction with model bodies (Rockwell, 1972; Rockwell and Knisely, 1979), vortical flows (Nagata *et al.*, 1979; Mochizuki and Yagi, 1982), or stratified flows (Moore and Long, 1971). Having become a standard measuring method, the technique has also been utilized for applied flow problems, e.g., model studies of the flow through aortic valves (van Steenhoven and van Dongen, 1979), or the hydrodynamics of ink jet printing (Hendriks, 1980).

2.3.3. Thymol Blue Method

It was mentioned in Section 2.1.2 that dye can be produced within a liquid by means of a pH indicator dissolved in the fluid. A number of dissolvable indicators are available that change color upon a specific change of the pH value (Table 2.2). Electrolysis allows for controlling the pH value and thereby for initiating the respective color change close to the electrodes. The solution becomes acidic near the anode and basic (alkaline) near the cathode. This is explained by the migration of positive hydrogen ions to the negative electrode (cathode); here they give up their charge and combine to form H_2 molecules. The solution, therefore, exhibits an excess of OH ions near the cathode and is thus alkaline. For this method it is important to have the rate of hydrogen production low

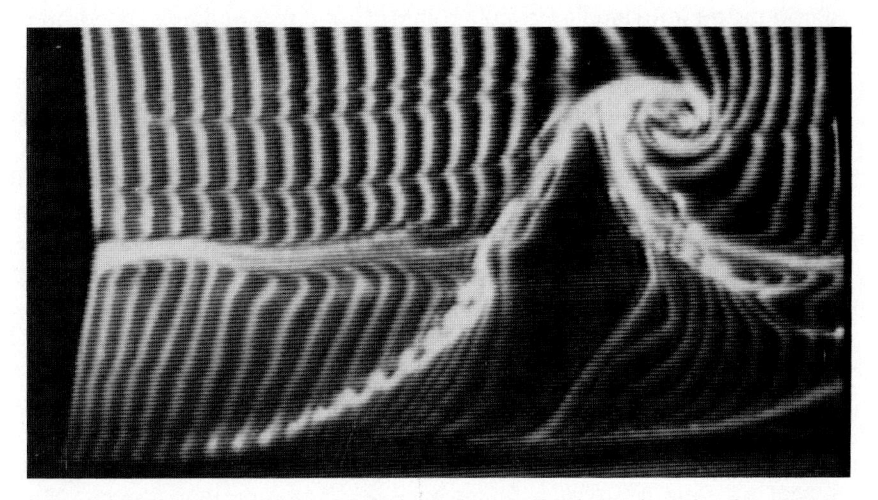

Fig. 2.37 Hydrogen-bubble time-line visualization of turbulent boundary layer pattern. Record photographed from TV screen. (Courtesy C. R. Smith, Lehigh University, Bethlehem, Pennsylvania.)

enough so as to avoid the formation of hydrogen bubbles. This requires the voltage applied between the electrodes to be low, a few volts only, in contrast to the conditions needed for the hydrogen bubble technique. The H_2 molecules can then be dissolved in the solution before they form gas bubbles.

The indicator most commonly used is thymol blue. Its use for flow visualization and the measurement of velocity profiles has been described by Baker (1966). The working fluid is an aqueous solution of thymol blue, which is orange yellow in an acidic environment ($pH < 8.0$) and turns its color to blue if the solution becomes alkaline ($pH > 9.6$; see Table 2.2). Two electrodes are introduced into the fluid; the cathode is a thin straight wire placed in the regime to be measured and perpendicular to the mean flow direction, as described already for the tellurium and hydrogen bubble technique. If one uses an acidic solution close to the point of becoming basic, the slight change of the pH-value near the cathode causes the solution in this regime to become blue. If the electric voltage is pulsed, a small column of dark blue solution is formed around the cathode wire, it will move with the flow and thereby be deformed according to the local velocity profile. By repetitive electric pulsing with a known pulse rate one obtains a series of time lines as discussed for the hydrogen bubble technique.

The test fluid is prepared by solving 0.01 to 0.04 wt. % of thymol blue in distilled water. In order to concentrate enough electrolytes in the flow one adds a few drops of NaOH, so that the color becomes deep blue. The solution is then titrated to its end point ($pH \cong 8.0$) by adding HCl drop by drop until the color turns yellow, that is, until one arrives on the acidic side of the end point. The life time of this solution can be several months; but the solution will decompose rapidly if one uses nondistilled water instead.

The shape of the electrode and the duration of the electric current determine the shape of the marked fluid element. A single platinum wire (diameter $\cong$ 0.01 mm) is used as the cathode for producing time lines. A copper plate or an equivalent network of wires may serve as the anode at some distance from the cathode. Typical voltages applied between these electrodes are in the order of 10 V at currents of about 10 mA or less. The error in measuring velocities, which arises from the presence of the wire wake has already been discussed for the hydrogen-bubble technique. Instead of making time lines visible, one may as well visualize streak lines, for example, with a partially isolated cathode wire and a steady electric current.

The optical contrast between the orange–yellow liquid and the blue dye produced at the cathode can be enhanced by employing a yellow-light illumination; a sodium discharge lamp is most appropriate for this purpose. Diffusion of the dye into the surrounding fluid increases with Reynolds number or mean flow velocity. This fact together with the limitation in the amount of blue dye producible per time restricts the application to maximum velocities of about 5 cm/sec.

The blue dye consists of thymol blue ions that remain in the state of solution. Hence, there are no density differences between the fluid elements marked by the dye and the rest of the fluid; the dye is "neutrally buoyant." The thymol blue technique is therefore used with preference for flow situations, which are dominated by mass forces (buoyancy, centrifugal forces), for example, convective flows (Hart, 1971; Vedhanayagam *et al.*, 1979; Incropera and Yaghoubi, 1980; Kimura and Bejan, 1980; Sparrow *et al.*, 1984; Gardner, 1985; Quraishi *et al.*, 1985), stratified flows (Buzyna and Veronis, 1971, see Fig. 2.38; Davies, 1972; Gershenfeld *et al.*, 1981), and flows in rotating systems, in many cases with the application to geophysical problems (Baker, 1967; Veronis and Yang, 1972; McCartney, 1975; Davies *et al.*, 1976; Maxworthy, 1977; Moll, 1978). Gerrard (1971) has measured velocity profiles in pulsating channel flow; Quraishi and Fahidi (1980) applied the thymol blue technique for studying the flow induced by electrolysis.

2.3.4. *Photochromic and Luminescent Dye*

The three former methods suffer from the presence of a disturbing electrode wire in the flow field of interest. In this section we discuss the possibility of producing the desired dye or tracer material along a straight line by irradiation of an appropriate light pulse. Thus, no disturbing mechanical element has to be introduced into the flow. The dye-producing light beam can be controlled externally in time and position. The fluid must contain elements, tracers or molecules, that react with the irradiated light to become visible fluid elements. It is desirable that such a reaction be reversible (i.e., that after a certain period the fluid regains its "invisible" pattern).

In the method described by Nakatani *et al.* (1975) the fluid, an aqueous solution of glycerin, is seeded with luminescent tracer particles, whose diameter is in the order of 10 μm. Luminescence of the tracers is excited by irradiation of UV light. The lifetime of the luminescence is between 40 and 1500 msec, depending on the specific tracer material. Since the wavelength of the emitted radiation (visible) is longer than that of the exciting radiation (UV), the emitted light is not reabsorbed by the surrounding tracer particles. The UV radiation is directed into the fluid along a straight

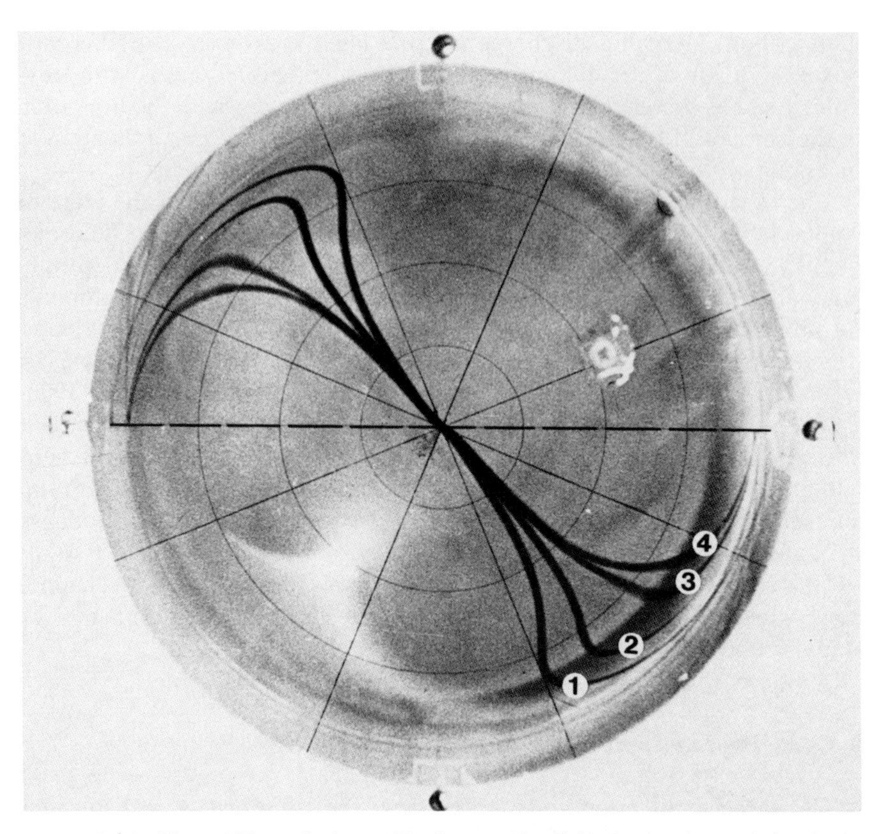

Fig. 2.38 Thymol blue velocity profiles in a rotating fluid after an abrupt change in the rotation of the container. Points 1–4 show different heights above the floor of the container. The position of the electrode wire is indicated by the dashed line. (From Buzyna and Veronis, 1971.)

line normal to the flow and in form of a short pulse of duration Δt_{ex}. Tracers within a certain volume element of the fluid are thereby excited, and the development of the shape of this volume element with time is sketched in Fig. 2.39. A normal Poiseuille flow has been assumed in this figure. Excitation begins at A and is stopped at B, so that the shaded region in B represents the volume of tracers, which had been excited during the period Δt_{ex}. Recording a photograph with an exposure time Δt_{rec} begins in C and is finished in D, so that the photographic exposure will furnish a pattern of the form shown in E. This sketch explains that the recorded pattern cannot appear uniformly luminescent, and it might also give some indication on the expected error when velocities are measured.

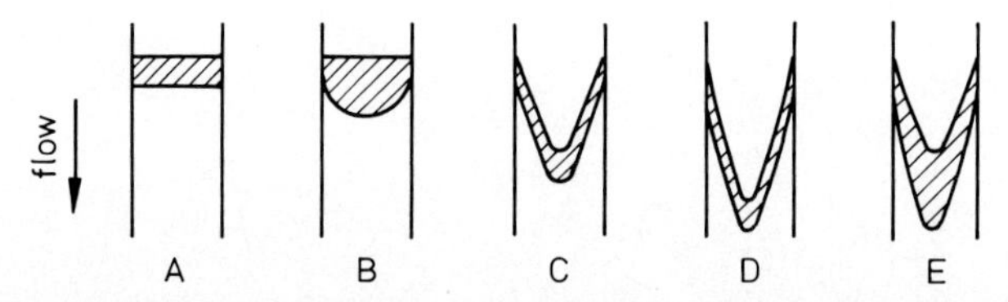

Fig. 2.39 Schematic development of a fluid element marked by the photochromic method in a Poiseuille flow. Laser excitation begins at time t(A) and ends at t(B) and photographic recording begins at t(C) and ends at t(D) so that the pattern E is recorded. (See Nakatani *et al.*, 1975.)

Nakatani *et al.* (1975) first used a xenon flash lamp for excitation, and they observed the visualized pattern in a two-dimensional flow with foreward scattering, probably due to limitations in intensity of the exciting light. Later, when using a N_2 laser for excitation, they were able to record the luminescent radiation in sideward direction, under 90° (Nakatani and Yamada, 1985). It was also shown (Nakatani *et al.*, 1977) that a fluid element of quadratic cross section can be excited, and that from the observed deformation of this element in the flow it is possible to conclude onto some turbulence characteristics.

Popovich and Hummel (1967) reported on the use of photochromic materials in liquid solutions for flow visualization. In the pertinent literature of photochemistry [e.g., Brown (1971)], "photochromism" is defined as the reversible transition of a chemical substance between two states exhibiting noticeably different absorption spectra. The possible absorption spectra of the two states, A and B, of such a substance is shown, in principle, in Fig. 2.40. State A has high absorption rates in the UV regime, while B is highly absorbing in the visible range of wavelengths. The substance is therefore visible (opaque) in state B and invisible (transparent) in A. The transition A → B is stimulated by (the absorbed) UV radiation, and substances are here of interest for which the reserved transition B → A is spontaneous and can be enhanced by thermal energy. The transition process can be described also by plotting the ratio of the number of molecules in state B, N_B, and the number of molecules in state A, N_A, as a function of time (Fig. 2.41). This ratio increases rapidly during the short period of incident UV radiation, and it decreases slowly during the reverse transition when the fluid regains its initial transparent state. The method consists of stimulating the transition A → B by means of a short pulse of UV radiation along a straight path and observing the motion of the generated opaque trace line. Since the reaction is reversible, the fluid could be used in a closed loop or for repeated experiments.

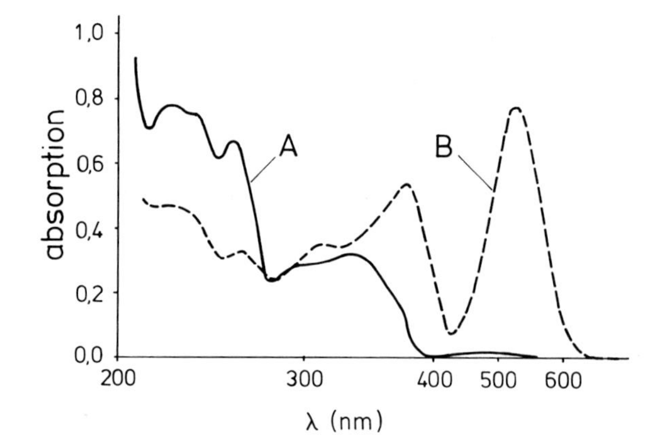

Fig. 2.40 Principle of absorption spectra of the two states, A and B, of a photochromic substance. Curve A shows the nonexcited state, high absorption in the UV and curve B the excited state, high absorption in the visible range.

Two different groups of photochromic substances have been used in the reported experiments, pyridines, e.g., 2–(2′,4′-dinitro-benzyl)–pyridine (Popovich and Hummel, 1967; Iribarne *et al.* 1972; Arunachalam *et al.,* 1972; Lavallee and Popovich, 1974), and spyrans, e.g., 1,3,3-trimethylindolino–6′–nitro-benzopyrylospyran (Humphrey *et al.,* 1974; Seeley *et al.,* 1975; Kondratas and Hummel, 1982). A disadvantage of these substances is that they are unsoluble in water. Solutions can be prepared with a number of organic liquids (e.g., ethanol, toluol, alcohol, kerosene). One is then limited by the properties of these fluids, and also the absorption characteristics of the substance (Fig. 2.40) can depend on the specific solvent. Concentration of the substance is between 0.01 and

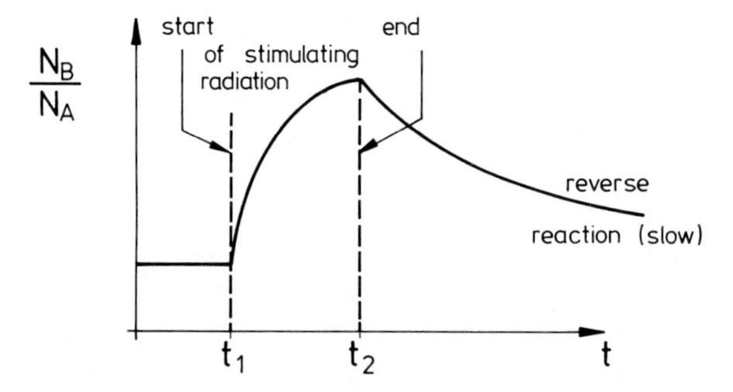

Fig. 2.41 Number of molecules in the nonexcited state of a photochromic substance, N_A, and number of molecules in the excited state, N_B, as function of time t.

0.2% by weight in the solvent. Choice of the concentration is a critical point for the technique. Too low a concentration results in a weakly visible pattern. Too high a concentration causes an unfavorable temperature increase in the fluid, because only a fraction of the incident energy is used for exciting the state B, and the rest is converted into heat, so that thermal convection overlapping the flow of interest may be initiated. The appropriate concentration is mainly determined by the type of available light source.

A mercury lamp or, as a better choice, a N_2 laser may serve as light sources. Short UV pulses can also be generated with a ruby laser equipped with a frequency doubler. A quartz window transparent for the UV radiation must be provided for the flow facility. Along its path the intensity of the light beam decreases, and it is necessary to guarantee the intensity everywhere above a certain minimum, which is needed for initiating the photochromic reaction. The visualized velocity profile in a channel flow is shown in Fig. 2.42. The "half-life" of the reversed reaction,

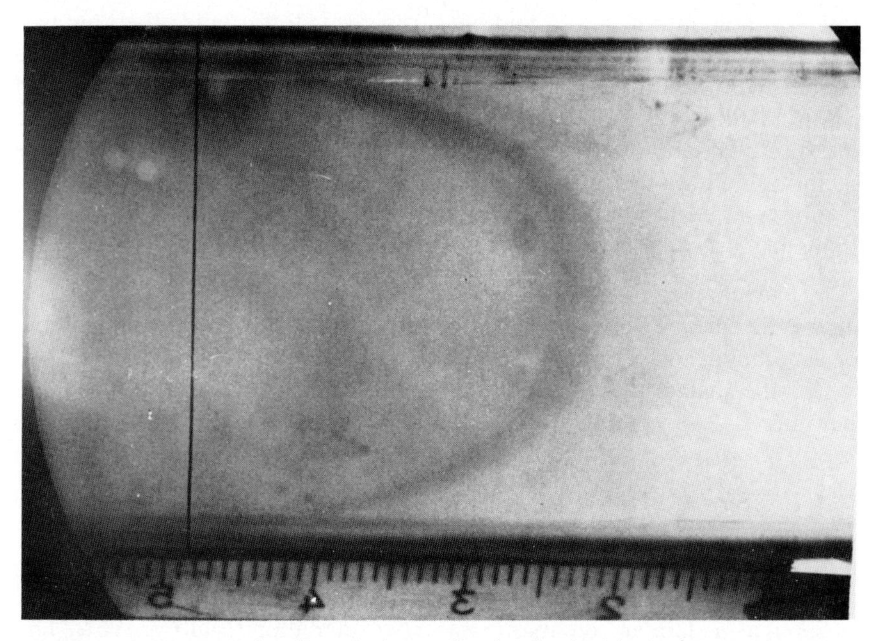

Fig. 2.42 Hagen–Poiseuille velocity profile visualized with photochromic reaction. The vertical wire on the right indicates the position of the laser beam exciting the reaction. The exposure was taken 1 sec after the laser was pulsed. The mean velocity is 1.32 cm/sec and the photochromic dye is TNSB and the fluid is C_6H_{14}. (From G. Fischer and W. Merzkirch, Ruhr–Universität Bochum.)

that is, the transition of the fluid to become again transparent, is of the order of a few seconds. This value can be compared with the 40 to 1500 msec duration of luminescence in the method discussed previously. Another difference between the two methods is that the photochromic dye in the fluid is a true solution, and therefore, not subject to buoyancy effects; whereas, the luminescent particles are buoyant tracers. The application of the photochromic method, as reported in the beforementioned references, was mostly in the near-wall region of boundary layer flows, where velocity profiles were measured with relatively short pathlengths of the incident UV light.

2.4. Visualization of Surface Flow

The interaction of a fluid flow with the surface of a solid body is a subject of great interest. Shear forces, pressure forces, and heating loads may be applied by the flow to the body. A useful means for estimating the rates of momentum, mass, and heat transfer is to visualize the flow pattern very close to the body surface. For this purpose the surface can be coated with the thin layer of a material, which upon the interaction with the fluid flow, develops a certain visible pattern. This pattern can be interpreted qualitatively and, in some cases, it is even possible to deduce quantitative data of the state of the flow close to the surface.

Three different interaction processes are known which deliver different kinds of information:

1. mechanical interaction (e.g., in the case of a surface oil film) visualizes the surface flow and may allow for a measurement of skin friction;
2. chemical interaction measures the mass transfer; and
3. thermal interaction measures the heat transfer or surface temperature.

2.4.1. Surface Flow Pattern

2.4.1.1. Oil Film Technique

Since many years of wind tunnel practice the surface oil film technique is taken as a standard technique for experimentation (Maltby, 1962). This technique serves for visualizing the flow pattern close to the surface of a solid body exposed to an air flow. The surface is coated with a specially prepared paint consisting of a suitable oil and a finely powdered pigment. Due to frictional forces the air stream carries the oil with it, and the remaining streaky deposit of the pigment gives an information on the direction of flow. Furthermore, it is believed that the observed pattern can indicate the positions of transition from laminar to turbulent flow in the wall boundary layer, and the positions of flow separation and reattachment. The question arised toward the interpretation and the reliability of these observations, because the presence of the oil film affects the boundary conditions for the air flow near the solid wall.

The only available analysis of the oil flow pattern has been developed by Squire (1962). The boundary conditions are that the oil velocity is equal to the air velocity in the boundary layer at the surface of the oil ($y = h$), that the viscous stresses in the oil and in the air are also equal at the oil–air surface, and that the oil is at rest at the body surface (see Fig. 2.43):

$$y = h: \qquad u_{\text{air}} = u_{\text{oil}}, \qquad u_{\text{air}} \frac{\partial u_{\text{air}}}{\partial y} = u_{\text{oil}} \frac{\partial u_{\text{oil}}}{\partial y},$$

$$y = 0: \qquad u_{\text{oil}} = 0.$$

The first condition means that the boundary layer of the air flow has a nonzero velocity at the oil surface. This velocity, u_{oil} for $y = h$, is deter-

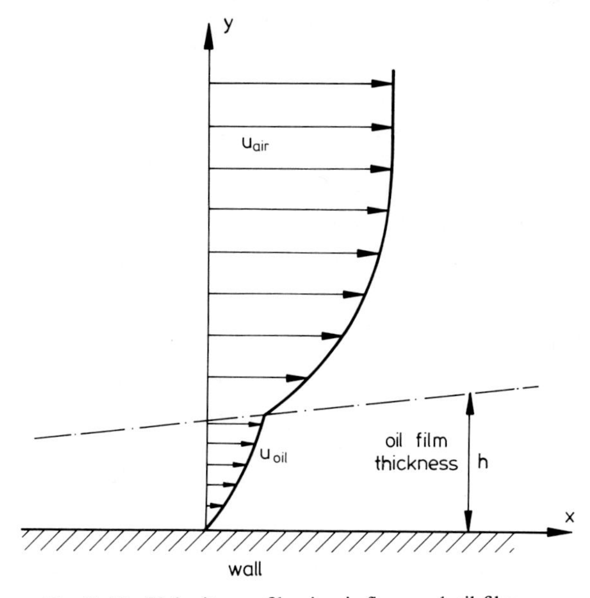

Fig. 2.43 Velocity profiles in air flow and oil film.

mined from the equations describing the slow viscous motion of the oil film, and it serves as a boundary condition for solving the (air) boundary layer equations:

$$u_{\text{oil}}(y = h) = \frac{\mu_{\text{air}}}{\mu_{\text{oil}}} \left(-\frac{h^2}{2} \frac{1}{\mu_{\text{air}}} \frac{\partial p}{\partial x} + h \left(\frac{\partial u_{\text{air}}}{\partial y} \right)_{y=h} \right). \tag{2.8}$$

The ratio $\lambda = \mu_{\text{air}}/\mu_{\text{oil}}$ is a small quantity, of the order 10^{-2} to 10^{-4}. The thickness of the oil film, h, is a function of time and assumed to be another small quantity. Therefore, u_{oil} is small and the boundary conditions are not changed too much compared to the situation if no oil were there. This assumption might not be true if the pressure gradient $\partial p/\partial x$ is large (e.g., near separation). This is a first indication that the oil flow pattern must be interpreted with care if the flow separates from the surface.

The curve of a "wall streamline" can be described by

$$\left(\frac{dx}{dz} \right)_{\text{air}} = \left(\frac{u_{\text{air}}}{w_{\text{air}}} \right)_{y \to 0} = \frac{(\partial u_{\text{air}}/\partial y)_{y=0}}{(\partial w_{\text{air}}/\partial y)_{y=0}}, \tag{2.9}$$

where x, z are coordinates in the surface of the wall, u, w the respective velocity components, and y is the coordinate normal to the wall. The question, whether the visible oil flow pattern is in agreement with the direction of the air "wall streamlines," can be answered with Squire's analysis. His result for the direction of an oil film streamline is

$$\left(\frac{dx}{dz} \right)_{\text{oil}} = \frac{(\partial u_{\text{air}}/\partial y)_{y=0}\mu_{\text{air}} + (\partial p/\partial x)[(y/2) - h]}{(\partial w_{\text{air}}/\partial y)_{y=0}\mu_{\text{air}} + (\partial p/\partial z)[(y/2) - h]}. \tag{2.10}$$

Hence, the direction of the oil flow depends on the wall shear stress (skin friction) of the air flow and on the pressure gradient. Generally, the pressure term is small compared with the skin friction, and the oil film pattern is a good representation of the actual "wall streamlines." Again, this assumption does not hold if the pressure term becomes large (e.g., near separation). Then, the oil flow is decelerated, the film thickness increases, the oil film may pile up to form a steep ramp, which affects the position where the flow separates from the wall. Of course, the same is true for reattachment.

Many instructions have been given in the literature on how to prepare an oil–pigment mixture, which is appropriate for specific test conditions (e.g., Maltby and Keating, 1962; Stanbrook, 1962; Settles and Teng, 1983). However, such instructions can only serve as guidelines for a first approach; it must be up to the experimentalist to develop the ideal mixture for his special purpose, and his result might be inadequate for the problems to be investigated in a different laboratory under different conditions. Therefore, all these instructions are given with a lack of precision, because precision appears to be inconsistent with the problem itself. The object is to prepare a mixture of such consistency that it will run easily enough under the given test conditions and leave behind the streaks of pigment indicating the direction of flow. Ideally, the mixture should not begin to run until the desired wind speed has been reached, and after a convenient time of running the pattern should be sufficiently dry to be unaffected after the air flow has been stopped.

The oils normally used are, in ascending order of viscosity: kerosene, light diesel oil, light transformer oil. For very low air speeds it might be useful to employ alcohol, or to adjust viscosity and surface tension of the oil by an additive. The high vapor-pressure characteristics of the standard oils are undesirable at very low test pressures as they usually occur in high-speed wind tunnels; in such cases vacuum-pump oil is an appropriate solution. The pigment in the mixture should provide a clear pattern against the background (i.e., the model surface). A white powder like titanium dioxide or china clay can be used on a dark model. Lampblack, a fine powder, is most suitable on a light model surface. It has very favorable mixing characteristics with oil and has become, therefore, a desirable pigment for wind-tunnel studies (e.g., Emery *et al.*, 1967; Sparrow *et al.*, 1979; Sparrow and Comb, 1983; Meznarsic and Gross, 1982). The dispersion of lampblack in oil may be improved by adding a few drops of oleic acid (Keener, 1983). A very bright surface pattern can be achieved with a fluorescent pigment (or dye), which should be illuminated with an UV source. Then, only the visualized surface flow pattern is seen, whereas, model and background remain invisible. This can be advantageous for complex model geometries (see, e.g., Gessner and Chan, 1983).

After the flow is turned off, the pigment pattern on the model surface can be examined, and for the purpose of recording, it can be photographed. If the model surface is curved, a problem exists in relating the curved surface pattern to a surface coordinate system. As a solution to this problem Sparrow and Comb (1983) cover the model surface with a white, plasticized, self-adhering contact paper and they apply the oil–pigment mixture to the paper. After the visualized pattern is formed, the paper is separated from the wall and laid flat for being photographed. For the same purpose, that is, for eliminating camera parallax when photographing the surface pattern, Settles and Teng (1983) apply transparent adhesive tapes to the model surface after the flow experiment. The very thin coating of pigment remaining on the surface is lifted off and preserved for taking a record.

A modification of this technique is to apply discrete dots of the indicator fluid on the surface of the test model, instead of providing a continuous coating or film over the whole surface. The oil streaks generated by the air flow give a better impression of the flow direction, and the method then is independent of the appropriate form of coagulations of the pigment (Peake *et al.*, 1972). Langston and Boyle (1982) describe a technique that is a synthesis of the oil-film and oil-dot method. A polyester drafting film is marked with a matrix of ink dots and then fastened to the body surface. The dots are made with a felt-tipped pen containing a water-insoluble ink. Before the air flow is started, the film is sprayed with a coating of oil ("oil of winter-green"). The ink dots dissolve and diffuse into the flowing oil film, so producing traces like the aforementioned oil streaks (Fig. 2.44). Since the oil is clear transparent, inks of different colors can be used.

A new development in wind tunnel testing are cryogenic wind tunnels, which can deliver high free-stream Reynolds numbers. Normal oil applied for surface flow visualization would freeze at the respective low temperature. What one needs instead is a cryogenic liquid with a low melting point. For such purposes, Kell (1978) has used propane (C_3H_8) that was mixed with a pigment to yield an opaque tracer liquid. Propane was cooled by liquid nitrogen and then pumped through holes in the model surface.

Different explanations have been given for the formation of the streaks, which become visible in the dry pigment layer remaining on the

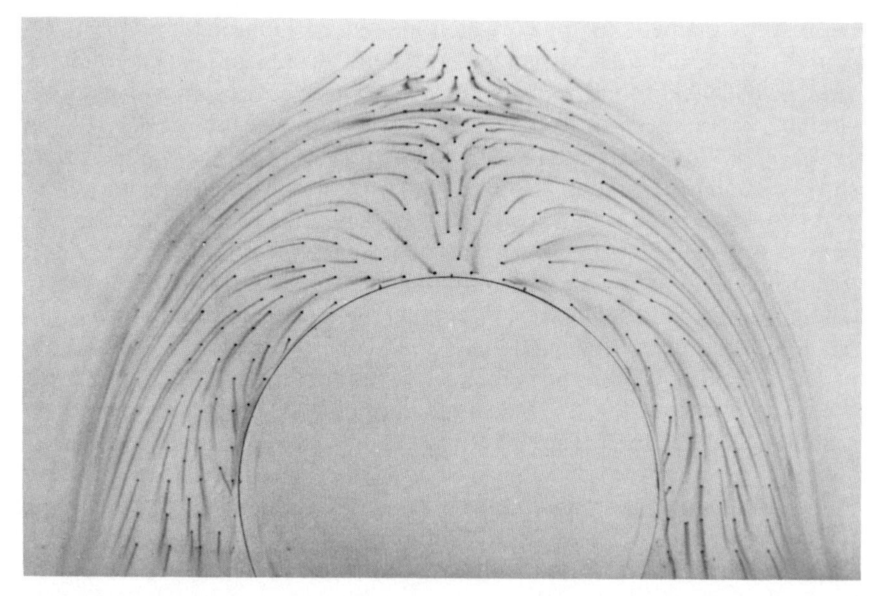

Fig. 2.44 Plan view of oil streaks in the flow around a circular cylinder mounted on a flat surface. (From Langston and Boyle, 1982.)

model surface, and which usually are interpreted as to indicate the direction of the flow. The streaks have been described as being caused by the wake behind coagulations of pigment particles. This would explain the use of oelic acid as an additive, which can control the formation and size of the coagulating flocks. Murai *et al.* (1982) have demonstrated that streaks develop in the oil film even without the presence of a pigment as a result of transverse instabilities.

The applications for aerodynamic testing are so numerous that only a few references can be listed, which are representative for many others. They range from low-speed testing (George, 1981) to high-speed configurations (Marvin *et al.*, 1972; Reding and Ericsson, 1982; Fig. 2.45) and include the visualization of flow separation due to the impingement of shock waves (Korkegi, 1976; Settles *et al.*, 1980).

The oil film technique could be applied to water flows, too. The ratio $\lambda = \mu_{water}/\mu_{oil}$ then might not be a very small quantity as in the case of air. Ishihara *et al.* (1982) investigated the separation of water flow from the surface of a circular cylinder with the axis normal to the main flow direction. The results with respect to the separation line were inconclusive.

2.4.1.2. *Skin Friction Measurements*

From the theoretical considerations in Section 2.4.1.1 one may conclude that it must be possible to relate the height of the oil film to the skin friction of the air flow on the surface, τ_w. It would be more exact to refer

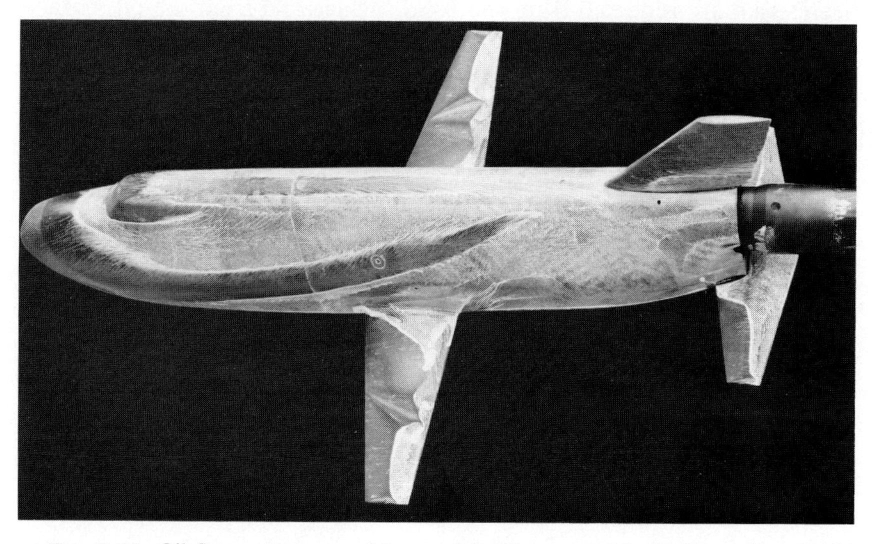

Fig. 2.45 Oil flow pattern on orbiter model. (Courtesy of L. H. Seegmiller, NASA Ames Research Center.)

to the skin friction at the air–oil interface; but we have learned already that the difference between this interface and the body surface is, in many cases, negligible. We suppose that the oil has been deposited on the body surface in the form of a small drop or along a straight line normal to the main flow direction. If the air flow is started at time $t = 0$, the thickness h of the oil film at a position x downstream of the film leading edge (the position of the original oil drop or oil line) and at time $t > 0$ is (Tanner and Blows, 1976):

$$h(x, t) = \frac{\mu}{t\tau_w^{1/2}} \int_0^x \frac{dx}{\tau_2^{1/2}}, \tag{2.11}$$

where μ is the dynamic viscosity of the oil. Tanner (1979, 1982) has demonstrated that the film thickness h can be measured by interferometric means (Fig. 2.46). Interference is generated between light rays reflecting from the surface of the oil film and from the surface of the solid body, respectively. If one takes an interferogram of the whole surface at a certain instant of time, $t = t_1$, the pattern of interference fringes carries information on the distribution of the film thickness, and therefore on the distribution of the skin friction or wall shear stress τ_W.

Such an interferogram can only be produced if the body surface is plane. For curved surfaces one may apply a modification of the method, which no longer delivers information on a whole field, but measures the oil film thickness at one point and as a function of time. Two focused laser beams that are reflected from the solid surface and from the oil surface, respectively, interfere with one another, as indicated in Fig. 2.46. Their interference is recorded with a photodiode, i.e., this photoelectronic re-

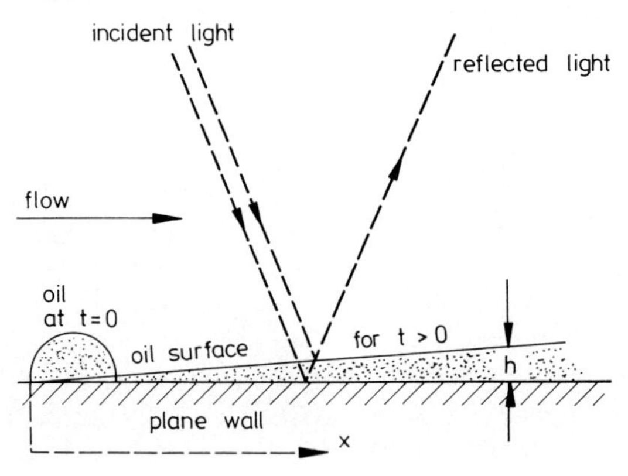

Fig. 2.46 Interferometric measurement of oil film thickness for determining skin friction. (After Tanner, 1979.)

ceiver records alternate bright and dark signals, according to the variation with time of the film thickness h at a fixed position x. Szodruch and Monson (1982) and Monson (1983) measured the wall shear stress in supersonic flow with this "skin friction probe."

As any oil film method, the technique will fail or produce unreliable results when applied to flow regimes close to separation or reattachment. Tanner (1977) showed that a measurement of the oil film thickness in a flow of known shear stress τ_w can serve for determining the viscosity of the oil.

2.4.1.3. *Wall Tufts*

A relatively simple means for getting an impression of the direction of air flow close to a solid wall is to attach one end of short tufts on the body surface. In laminar flow these tufts may well indicate the local flow direction. When the air flow becomes unsteady or turbulent, however, the tufts exhibit a certain unsteady motion, and this may be taken as an indication that the wall boundary has become turbulent. A more violent motion of the tufts, or a tendency to lift from the surface, may indicate a separated flow regime.

The choice of tuft size and material depends on the flow conditions and the size of the model to be tested. Tufts made of ordinary yarn and several centimeters long have been used for testing full-scale car models and even on full-scale airplanes in free flight. It is obvious that tufts of this size and the corrugation caused by the glue or the device fixing the tufts affects the surface conditions for the flow. In order to minimize such interference with the flow Crowder (1982) has used "minitufts" made from thin fluorescent nylon monofilament. These tufts have a diameter of about 20 μm, and their visibility is enhanced by observing or photographing them under UV illumination (Fig. 2.47).

2.4.2. *Surface Mass Transfer*

The mass transfer from a solid surface to a fluid flow or vice versa is itself subject of great interest. Additionally it is possible to conclude from the known rate of mass transfer onto the heat transfer, if Reynolds analogy is applicable. A number of methods for visualizing and measuring surface mass transfer are known. The surface is coated with a material, which changes shape or color as a function of the mass transfer rate. In case of the naphthalene sublimation technique (Tien and Sparrow, 1979; Sparrow *et al.*, 1983) the solid surface is coated with a layer of dense naphthalene, the outer surface of the layer being smooth (plane or circularly round) to a high degree of precision. The naphthalene is removed from

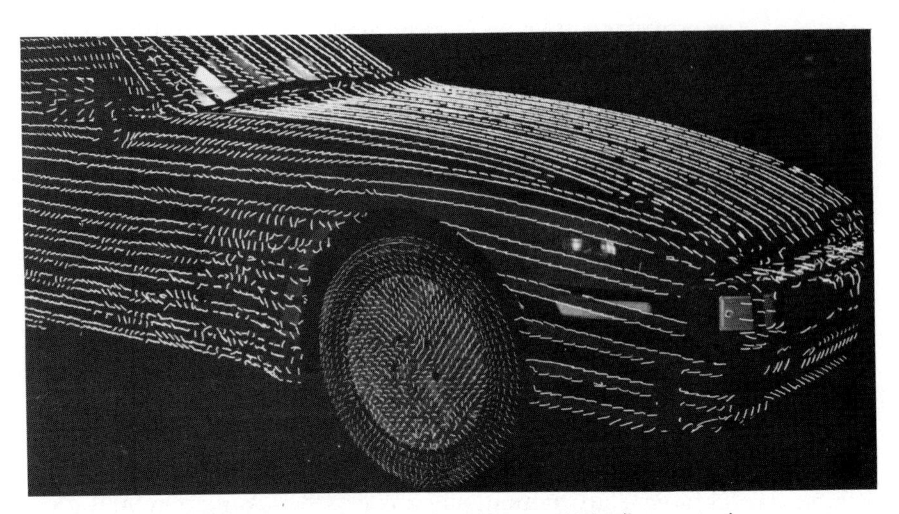

Fig. 2.47 Fluorescent minitufts on car moving at 160 km/h past stationary camera. (Courtesy J. P. Crowder, X-Aero Company, Seattle, Washington.)

the body because of sublimation evaporation into the air flow, and the resulting structure of the originally smooth surface visualizes the areas of different rates of sublimation or mass transfer. A measurement of the naphthalene profile enables one to obtain quantitative values of the local mass transfer coefficient.

Ablation, a special kind of sublimation, is the cause for the surface relief that can be observed on bodies that have been exposed to high-enthalpy, high-speed flows. Different hypotheses have been presented for the formation of the mostly observed cross-hatched pattern of ablating bodies (Swigart, 1974; see Fig. 2.48). It is likely that both the temperature and pressure field contribute to the pattern. Analogous to the use of sublimation methods in gas flows is the application of plaster of paris ($CaSO_4$) in water flows (Allen, 1966). This material is slightly soluble in water, and streaks appear on the surface according to the rate of the water flow near the surface.

A more discernible visualization is achieved with methods in which the mass transfer initiates a chemical reaction resulting in a color change on the body surface. In the technique described by Kottke *et al.* (1977) and Kottke (1982), the wall surface is coated with a thin wet layer (filter paper or gel) containing an aqueous solution of $MnCl_2$ and H_2O_2. A reacting gas, NH_3, is added in form of a short pulse and at low concentration to the main air stream. The gas, NH_3, is absorbed, according to the local partial concentration differences, by the wet layer, and a reaction takes place in

Fig. 2.48 Cross-hatched ablation pattern on Teflon test model. (From Nachtsheim and Larson, 1971.)

which MnO_2 is formed as an end product. The observed color intensity of MnO_2 is a measure of the local mass transfer rate, and this value can be measured quantitatively by photometric means (Fig. 2.49). An interesting result of respective investigations is that the measured position of maximum mass transfer in a reattaching flow is not consistent with the line of reattachment as observed, e.g., with the oil film technique. This is just another indication for the care, which is needed in the interpretation of visualized patterns of separated flows.

A similar method, also employing NH_3 as a reacting gas added to the main air flow, has been applied by Sadeh *et al.* (1982). The body wall is coated with a dry-surface layer containing "Congo Red" as a pH indicator. The color of the layer will change according to the local mass transfer rate. Injecting NH_3 into the recirculating base flow behind a body allows for visualizing the separation line. Peterson and Fitzgerald (1980) use a fluorescent spray as the surface coating. Fluorescence is excited by illumination with blue light. Oxygen quenching of the fluorescence causes those portions of the surface to remain dark, which experience a strong mass transfer rate of oxygen from the outer flow.

Fig. 2.49 Visualization of mass transfer to the surface of a car model in a wind tunnel. White areas indicate separated flow regimes with low mass transfer rate. (From Kottke, 1982. Published by Hemisphere Publishing Corporation.)

2.4.3. Surface Heat Transfer Visualization

Aerodynamic heating loads can be visualized by means of coatings whose pattern or color reacts to changes of the surface temperature. As will be shown later, the derivation of aerodynamic heat transfer coefficients requires the recording of the time history of a number of isotherms on the surface. For this purpose, temperature-sensitive paints have been used since many years of high-speed wind-tunnel testing (Cérésuela *et al.*, 1965; Kafka *et al.*, 1965; Pontézière and Bétremieux, 1967; Creel and Hunt, 1972). These paints are composed of several components each of which undergoes a visible change of its internal structure or phase at a specific temperature. The edge between two colors is a curve of constant surface temperature. If the model is made of a material with low thermal conductivity, the indicated temperature may be taken as the adiabatic wall temperature. The edges between colors move on the body surface in downstream direction with increasing time, until the whole body surface might have reached a uniform temperature.

The most frequently used temperature sensitive paints exhibit four to five changes in color as a function of temperature. The characteristics of Thermocolor (trade name) show that the color changes also depend on the time during which the paint has been exposed to a certain temperature (Fig. 2.50). The sequence of colors for Detectotemp (trade name) is pink–light blue–yellow–black–olive green. The range of sensitivity is included in Fig. 2.50. The models coated with the paint must be injected quickly

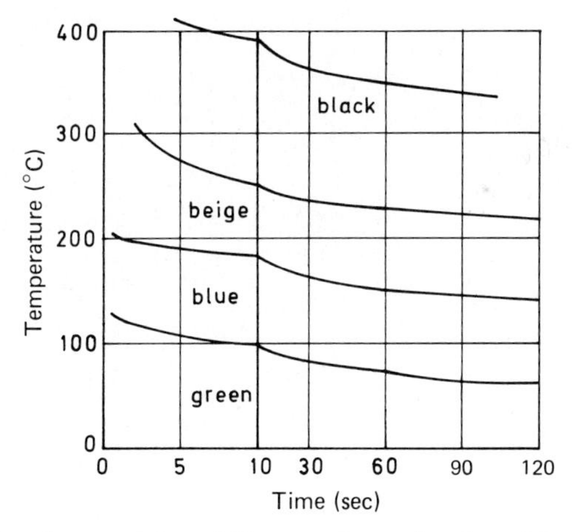

Fig. 2.50 Color characteristics of Thermocolor as function of temperature and time during which paint was exposed to heat. (From Cérésuela *et al.*, 1965.)

into the wind tunnel. The heat transfer to the surface can be determined by recording the time history of the color changes. Besides applying a respective heat transfer analysis (see below), it is also possible to calibrate the observed color changes for a quantitative evaluation. Such a calibration can be performed by applying the paint to a sphere and exposing the sphere to the same wind tunnel flow as the test model. The sphere is a body whose aerodynamic heat transfer rates are well known as a function of position on the surface. From the development of the color changes on the spherical surface (Fig. 2.51), and from the known heat transfer coefficients of the sphere, one obtains a set of calibration curves (Cérésuela *et al.*, 1965).

The edge between two colors of the paint is to a certain degree diffusive, and this behavior reduces the local resolution of the mapping. More accurate in this respect is the use of a one-component paint ("phase change coating"), which changes from solid to liquid at a precisely known surface temperature (Jones and Hunt, 1966; Segletes, 1975; Drummond *et al.*, 1976; Maegley and Carroll, 1982). Wax can be the major component of such coating. The observed melt line is an isotherm. Keyes (1976) reports that the coating can be blown off due to high shear rates in the flow, and this phenomenon could be misinterpreted to be a phase change.

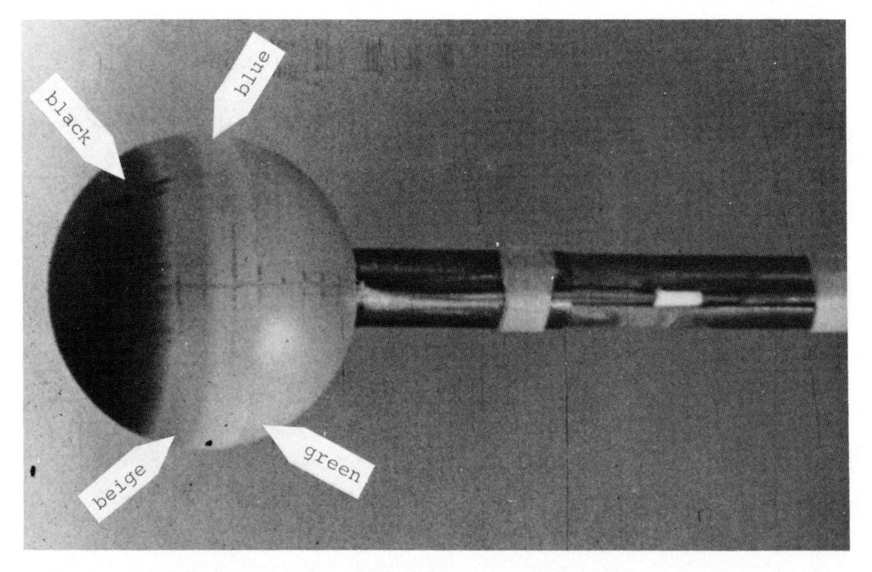

Fig. 2.51 Pattern of thermosensitive paint on a sphere after 3 sec of exposure to a hypersonic, high-enthalpy flow; Mach number $M = 10$, stagnation temperature = 1000 K. (From Cérésuela *et al.*, 1965.)

In both cases, temperature sensitive paint and phase change coating, the process causing the visible color change is not reversible, so that this coating of the surface can be used only once.

A reversible color change as a function of the surface temperature can be obtained with a coating of liquid crystals. Klein (1968a,b) gave a first description of the use of liquid crystals for the study of aerodynamic surface heating. Temperature changes in such crystals disturb the intermolecular forces and cause a shift in the molecular structure, thus causing a shift in wavelength of the light scattered from the crystals. An extensive literature is available on liquid crystals and their applications (e.g., Chandrasekhar, 1977). Among the various types, cholesteric liquid crystals have been proven to be most suitable for studies in fluid mechanics and convective heat transfer. Unlike temperature sensitive paints, cholesteric liquid crystals react to changes of temperature only within a few degrees Kelvin. The crystals are commercially available, for a great range of both bandwidth and mean temperature at which they are sensitive. Within the bandwidth of their sensitivity, up to ten color hues may be discriminated by the eye. The visible edge between two colors is a curve of constant temperature.

The dependence of indicated color on temperature is given in a diagram showing the spectral behavior of the liquid crystal (Fig. 2.52). It might be necessary to recalibrate the crystals if they are used in an environment different from that for which the original calibration had been performed.

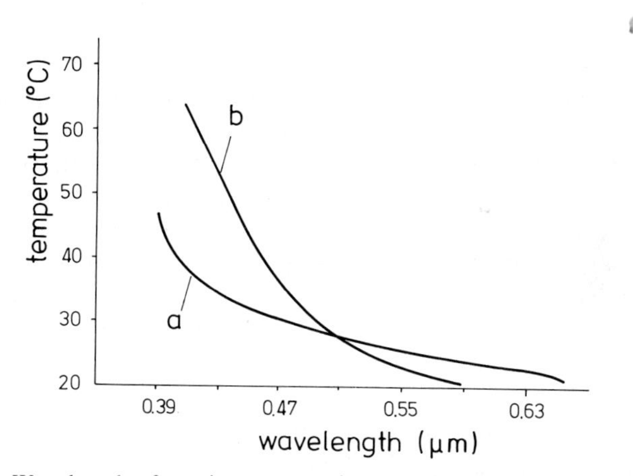

Fig. 2.52 Wavelength of maximum scattering as a function of temperature for two cholesteric liquid crystals: (a) 20% cholesteryl acetate and 80% cholesteryl nonanoate; (b) 20% cholesteryl methyl carbonate and 80% cholesteryl nonanoate. (After Chandrasekhar, 1977.)

Simonich and Moffat (1982) describe such a calibration procedure for liquid crystals used on a surface, which is exposed to a water flow. A thin sheet of liquid crystals on a solid surface might be affected by shear forces, or if they are used in water, by contamination. In order to protect the crystals, they can be made available in the form of sheets consisting of the liquid crystals laid on plastic material and covered by a transparent protecting layer (Hippensteele *et al.*, 1983; Kitamura *et al.*, 1985), or they are "microencapsulated," that is, enclosed in plastic spheres of 10 to 50 μm in diameter (Schöler, 1978; Ogden and Hendricks, 1984). Dispersing the encapsulated liquid crystals in water gives an emulsion, which can be brushed or sprayed on the model surface.

Cholesteric liquid crystals have been applied for surface thermography in heat transfer studies if the temperature does not exceed a value of about 500 K (Cooper *et al.*, 1975; Goldstein and Timmers, 1982; Kang *et al.*, 1982), and for visualizing flow contours like longitudinal vortices or eddy structures (Kitamura *et al.*, 1985; Schöler, 1978; Simonich and Moffat, 1982). The derivation of heat transfer coefficients is described below.

The surface of a solid body emits infrared (IR) radiation whose intensity is a function of the surface temperature. This radiation can be received with an IR camera, which in connection with appropriate computer hardware and software, converts the signal into a visual pattern. Several successful attempts have been made for mapping the surface temperature distribution of models in wind tunnels or in other flow environments by means of an IR camera (Boylan *et al.*, 1978; Bandettini and Peake, 1979; Cerisier *et al.*, 1982). Most of the reported experiments have been performed with an IR camera, which is commercially available; it employs a detector sensitive to the wavelength range between 3 and 5.8 μm. The detector, which must be cooled by liquid nitrogen, produces an electric signal proportional to the received total energy (Gauffre and Fontanella, 1980). The total radiant energy W emitted by the surface under investigation is, according to the Stefan–Boltzmann law for a graybody, proportional to the 4th power of the absolute temperature: $W = \varepsilon\sigma T^4$, σ being the Stefan–Boltzmann constant. The surface emissivity $\varepsilon \leqq 1$ is a function of wavelength and camera viewing angle and must be determined experimentally. Boylan *et al.* (1978) describe the use of a blackbody radiator of known temperature for calibration of the system. It is advisable to paint the model surface black.

The camera system records and displays the surface temperature pattern in real time. Figure 2.53 schematically describes the application of such a system to wind tunnel experiments. The viewing windows must be made of special glass (e.g., zinc sulfide) transparent for the IR radiation. The signal pattern recorded by the camera can be digitized, processed and

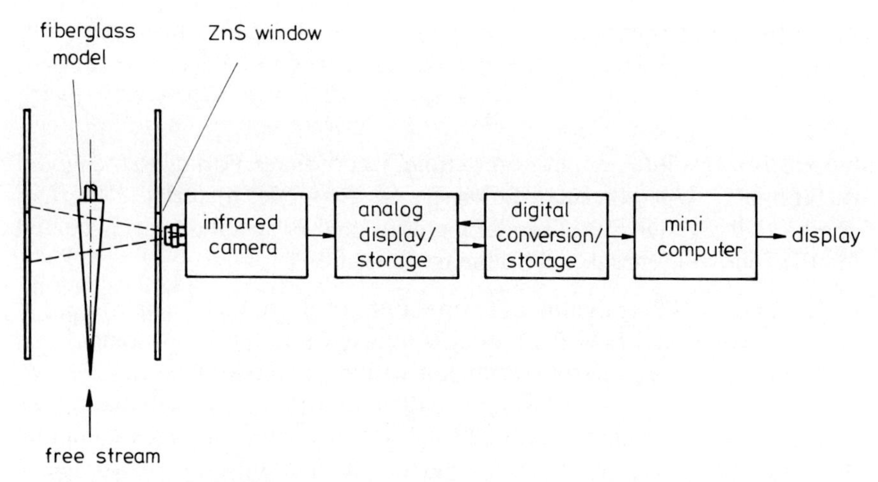

Fig. 2.53 Schematic representation of infrared camera system and instrumentation. (After Bandettini and Peake, 1979.)

displayed on either color or black-and-white monitor. Contours of constant surface temperature can be shown, and the calculation of local values of the heat transfer coefficient follows the procedure of Jones and Hunt (1966).

Table 2.5 summarizes the characteristics of the four possible coatings or methods, which can display surface isotherms. Jones and Hunt (1966) have devised a way for determining the aerodynamic heat transfer coefficient h, once the pattern of the isotherms is recorded as a function of time (e.g., by taking a movie film). The method is based on the transient, one-dimensional analysis of heat flow from the model surface into the body.

TABLE 2.5

Methods for Surface Temperature Mapping

Coating/method	Isotherms	Temperature range
Phase change paint	T_1	One temperature value
Temperature sensitive paint	$T_1, T_2, \ldots, T_n$	Wide range ΔT
IR camera	$T_1, T_2, \ldots, T_n$	Medium range ΔT
Liquid crystals	$T_1, T_2, \ldots, T_n$	Narrow range ΔT

The differential equation

$$\frac{\partial T}{\partial t} = \alpha \frac{\partial^2 T}{\partial y^2}, \tag{2.12}$$

(where T is absolute temperature, t time, y coordinate normal to the model surface, $\alpha = k/\rho c_p$ thermal diffusivity of the model material, k thermal conductivity, ρ density of model material, and c_p specific heat) is solved for the following initial and boundary conditions:

1. $T(y, 0) = T_i$ is the initial temperature of the model. In practice, $t = 0$ is the instant at which the model is injected into the wind tunnel.
2. $T(\infty, t) = T_i$: thermal diffusion within the model material is slow.
3. $[\partial T(0, t)/\partial x] = (h/k)(T_{aw} - T(0, t))$, with T_{aw} the adiabatic wall temperature and h aerodynamic heat transfer coefficient. This condition describes the development of the surface temperature (at $y = 0$) due to aerodynamic heating.

With a number of additional simplifying assumptions, Jones and Hunt develop the following solution for the heat flow equation [Eq. (2.12)]:

$$T^* = 1 - \exp(\beta^2)\text{erfc}\ \beta \tag{2.13}$$

with the abbreviations

$$T^* = (T_{ind} - T_i)/(T_{aw} - T_i), \qquad \beta = (h/k)\sqrt{\alpha t},$$

$$erfc\ \beta = \frac{2}{\pi}\int_{\beta}^{\infty} e^{-\lambda^2}\, d\lambda,$$

where T_{ind} is the temperature of an isotherm indicated by the specific coating. The determination of h is as follows. With given T_{ind} for a specific indicator one determines the value of T^* and, from a plot T^* versus β (see Jones and Hunt, 1966) the corresponding value of β. With known values k, α, and measured time t (for the isotherm to appear at a specific position on the surface, as measured from the movie film), one determines $h = k \cdot \beta/\sqrt{\alpha t}$. The time required for the isotherms to move over the model surface should be large compared with the time necessary for establishing a steady flow around the model, and it must be short compared with the thermal diffusion time of the model material. Segletes (1975) has pointed out that Jones and Hunt's one-dimensional heat transfer assumption, in some cases, may lead to large errors. He presents correction factors for improving the results of the one-dimensional analysis when applied to particular corner flow situations.

For measuring the local surface heat transfer to a model body in water flow, Simonich and Moffat (1982) report on a method which does not depend on taking time records of the patterns of isotherms. They use a sheet of liquid crystals beneath a transparent sheet of gold whose thickness is a few angstroms. The gold sheet is electrically heated and serves as a constant temperature heat source. The temperature and thereby the color of the liquid crystals is determined by the temperature difference between the gold sheet and the outside surface of the coating.

References

Sections 2.1.1 and 2.1.2

Breidenthal, R. (1979). Chemically reacting, turbulent shear layer. *AIAA J.* **17,** 310–311.

Breidenthal, R. (1981). Structure in turbulent mixing layers and wakes using a chemical reaction. *J. Fluid Mech.* **109,** 1–24.

Breidenthal, R. (1982). Reacting flow visualization of a turbulent shear layer and wake. *In* "Flow Visualization II" (W. Merzkirch, ed.), pp. 689–693. Hemisphere, Washington, D.C.

Clayton, B. R., and Massey, B. S. (1967). Flow visualization in water: a review of techniques. *J. Sci. Instrum.* **44,** 2–11.

Csanady, G. T. (1963). Turbulent diffusion in Lake Huron. *J. Fluid Mech.* **17,** 360–384.

Danckwerts, P. V., and Wilson, R. A. M. (1963). Flow visualization by means of a time-reaction. *J. Fluid Mech.* **16,** 412–416.

Davis, E. J., and Choi, C. K. (1977). Cellular convection with liquid-film flow. *J. Fluid Mech.* **81,** 565–592.

Deardorff, J. W., Willis, G. E., and Stockton, B. H. (1980). Laboratory studies of the entrainment zone of a convectively mixed layer. *J. Fluid Mech.* **100,** 41–64.

Delisi, D. P., and Orlanski, I. (1975). On the role of density jumps in the reflexion and breaking of internal gravity waves. *J. Fluid Mech.* **69,** 445–464.

Denbigh, K. G., Dombrowski, N., Kisiel, A. J., and Place, E. R. (1962). The use of the "time reaction" in residence time studies. *Chem. Eng. Sci.* **17,** 573.

Donohue, G. L., Tiederman, W. G., and Reischman, M. M. (1972). Flow visualization of the near-wall region in a drag-reducing channel flow. *J. Fluid Mech.* **56,** 559–575.

Dumas, R., Domptail, C., and Daien, E. (1982). Hydrodynamic visualization of some turbulent flow structures. *In* "Flow Visualization II" (W. Merzkirch, ed.), pp. 393–397. Hemisphere, Washington, D.C.

Erickson, G. E. (1979). Water tunnel flow visualization: Insight into complex three-dimensional flow fields. *AIAA Pap.* **79-1530.**

Faler, J. H., and Leibovich, S. (1977). Disrupted states of vortex flow and vortex breakdown. *Phys. Fluids* **20,** 1385–1400.

Fiechter, M. (1969). Wirbelsysteme schlanker Rotationskörper. *Jahrb. 1969, Dtsch. Ges. Luft- Raumfahrt (DGLR)* pp. 77–85.

Gad-el-Hak, M., Blackwelder, R. F., and Riley, J. J. (1981). On the growth of turbulent regions in laminar boundary layers. *J. Fluid Mech.* **110,** 73–95.

Gaster, M. (1969). Vortex shedding from slender cones at low Reynolds number. *J. Fluid Mech.* **38,** 565–576.

Hama, F. R. (1962). Streaklines in a perturbed shear flow. *Phys. Fluids* **5,** 644–650.
Han, T., and Patel, V. C. (1979). Flow separation on a spheroid at incidence. *J. Fluid Mech.* **92,** 643–657.
Hide, R., and Titman, C. W. (1967). Detached shear layers in a rotating fluid. *J. Fluid Mech.* **29,** 39–60.
Hung, T.-K., and Brown, T. D. (1976). Solid-particle motion in two-dimensional peristaltic flows. *J. Fluid Mech.* **73,** 77–96.
Hunt, J. C. R., and Snyder, W. H. (1980). Experiments on stably and neutrally stratified flow over a model three-dimensional hill. *J. Fluid Mech.* **96,** 671–704.
Kang, I. S., and Chang, H. N. (1982). The effect of turbulence promoters on mass transfer—Numerical analysis and flow visualization. *Int. J. Heat Mass Transfer* **25,** 1167–1181.
Kotas, T. J. (1977). Streamline pattern in a confined vortex flow. *J. Mech. Eng. Sci.* **19,** 38–41.
Lynch, W. H., and Brown, A. E. (1974). Flow measurement by fluorescing dye dilution techniques. *In* "Flow—Its Measurement and Control in Science and Industry" (R. B. Dowdell, ed.), pp. 781–785. ISA, Pittsburgh, Pennsylvania.
Martin, B. W., and Lockwood, F. C. (1964). Entry effects in the open thermosyphon. *J. Fluid Mech.* **19,** 246–256.
Masliyah, J. H. (1972). Steady wakes behind oblate spheroids: Flow visualization. *Phys. Fluids* **15,** 1144–1146.
Masliyah, J. H. (1980). On laminar flow in curved semicircular ducts. *J. Fluid Mech.* **99,** 469–479.
Maxworthy, T. (1972). The structure and stability of vortex rings. *J. Fluid Mech.* **51,** 15–32.
McNaughton, K. J., and Sinclair, C. G. (1966). Submerged jets in short cylindrical flow vessels. *J. Fluid Mech.* **25,** 367–375.
Offen, G. R., and Kline, S. J. (1974). Combined dye-streak and hydrogen-bubble visual observations of a turbulent boundary layer. *J. Fluid Mech.* **62,** 223–239.
Pullin, D. I., and Perry, A. E. (1980). Some flow visualization experiments on the starting vortex. *J. Fluid Mech.* **97,** 239–255.
Quraishi, M. S., and Fahidy, T. Z. (1982). A flow visualization technique using analytical indicators: theory and some applications. *Chem. Eng. Sci.* **37,** 775–780.
Sarpkaya, T. (1971). On stationary and travelling vortex breakdowns. *J. Fluid Mech.* **45,** 545–559.
Simpson, J. E. (1972). Effects of the lower boundary on the head of a gravity current. *J. Fluid Mech.* **53,** 759–768.
Sullivan, P. J. (1971). Longitudinal dispersion within a two-dimensional turbulent shear flow. *J. Fluid Mech.* **49,** 551–576.
Thomas, A. S. W., and Cornelius, K. C. (1982). Investigation of a laminar boundary-layer suction slot. *AIAA J.* **20,** 790–796.
Utami, T., and Ueno, T. (1984). Visualization and picture processing of turbulent flow. *Exp. Fluids* **2,** 25–32.
Werlé, H. (1960). Étude effectuée à la cuve à huile et au tunnel hydrodynamique à visualisation de l'O.N.E.R.A. *Rech. Aeronaut.* No. 79, 9–26.
Werlé, H. (1973). Hydrodynamic flow visualization. *Annu. Rev. Fluid Mech.* **5,** 361–382.
Werlé, H. (1979). Application of hydrodynamic visualization to the study of low speed flow around a delta wing aircraft. *In* "Flow Visualization" (T. Asanuma, ed.), pp. 111–116. Hemisphere, Washington, D.C.
Werlé, H. (1980). Transition et décollement: visualisations au tunnel hydrodynamique de l'ONERA. *Rech. Aerosp.* No. 1980-5, 331–345.
Werlé, H. (1982). Hydrodynamic visualization on streamlined bodies of vortex flows partic-

ular to high angles of attack. *In* "Flow Visualization II" (W. Merzkirch, ed.), pp. 373–378. Hemisphere, Washington, D.C.
Werlé, H., and Gallon, M. (1976). Étude par visualisations hydrodynamiques de divers procédés de contrôle d'écoulements décollés. *Rech. Aerosp.* No. 1976-2, 75–94.
Withjack, E. M., and Chen, C. F. (1974). An experimental study of Couette instability of stratified fluids. *J. Fluid Mech.* **66,** 725–737.
Yeheskel, J., and Kehat, E. (1971). Wakes of vertical and horizontal assemblages of drops. *Chem. Eng. Sci.* **26,** 2037–2042.

Sections 2.1.3 and 2.1.4

Ahmed, S. R. (1983). Influence of base slant on the wake structure and drag of road vehicles. *J. Fluids Eng.* **105,** 429–434.
Baker, C. J. (1979). The laminar horseshoe vortex. *J. Fluid Mech.* **95,** 347–367.
Batill, S. M., and Mueller, T. J. (1981). Visualization of transition in the flow over an airfoil using the smoke-wire technique. *AIAA J.* **19,** 340–345.
Batill, S. M., Nelson, R. C., Mueller, T. J., and Wells, W. C. (1982). Smoke visualization at transonic and supersonic Mach numbers. *AIAA Pap.* **82-0188.**
Becker, H. A., Hottel, H. C., and Williams, G. C. (1967). On the light-scatter technique for the study of turbulence and mixing. *J. Fluid Mech.* **30,** 259–284.
Bird, J. D. (1952). Visualization of flow fields by use of a tuft grid technique. *J. Aeronaut. Sci.* **19,** 481–485.
Bisplinghoff, R. L., Coffin, J. B., and Haldeman, C. W. (1976). Water fog generation system for subsonic flow visualization. *AIAA J.* **14,** 1133–1135.
Bouchez, J. P., and Goldstein, R. J. (1975). Impingement cooling from a circular jet in a cross flow. *Int. J. Heat Mass Transfer* **18,** 719–730.
Bradshaw, P. (1970). "Experimental Fluid Mechanics," 2nd Ed. Pergamon, Oxford.
Brown, F. N. M. (1953). A photographic technique for the mensuration and evaluation of aerodynamic patterns. *Photogr. Eng.* **4,** 146–156.
Cashdollar, K. L., Lee, C. K., and Singer, J. M. (1979). Three-wavelength light transmission technique to measure smoke particle size and concentration. *Appl. Opt.* **18,** 1763–1769.
Cenkner, A. A., Jr., and Driscoll, R. J. (1982). Laser-induced fluorescence visualization on supersonic mixing nozzles that employ gas-trips. *AIAA J.* **20,** 812–819.
Cherdron, W., Durst, F., and Whitelaw, J. H. (1978). Asymmetric flow and instabilities in symmetric ducts with sudden expansions. *J. Fluid Mech.* **84,** 13–31.
Corke, T., Koga, D., Drubka, R., and Nagib, H. (1977). A new technique for introducing controlled sheets of smoke streaklines in wind tunnels. *ICIASF '77 Rec.,* 74–80.
Dewey, J. M. (1971). The properties of a blast wave obtained from an analysis of the particle trajectories. *Proc. R. Soc. London, Ser. A* **324,** 275–299.
Dewey, J. M., and Walker, D. K. (1973). The particle trajectories in a two-dimensional shock tube flow observed with a double-pass laser schlieren system. *In* "Recent Developments in Shock Tube Research" (D. Bershader, W. Griffith, eds.), 285–293. Stanford Univ. Press, Stanford, California.
Farukhi, N. M. (1983). Laser generated smoke for fluid flow visualization. *J. Heat Transfer* **105,** 413–414.
Freymuth, P., Bank, W., and Palmer, M. (1983). Visualization of accelerating flow around an airfoil at high angles of attack. *Z. Flugwiss. Weltraumforsch.* **7,** 392–400.

Gad-el-Hak, M., and Morton, J. B. (1979). Experiments on the diffusion of smoke in isotropic turbulent flow. *AIAA J.* **17,** 558–562.
Gersten, K. (1956). Untersuchungen über den Abwind hinter Deltaflügeln bei inkompressibler Strömung. *Jahrb. Wiss. Ges. Luft-Raumfahrt* **1955,** 151–160.
Goddard, V. P., McLaughlin, J. A., and Brown, F. N. M. (1959). Visual supersonic flow patterns by means of smoke lines. *J. Aerosp. Sci.* **26,** 761–762.
Griffin, O. M., and Votaw, C. W. (1973). The use of aerosols for the visualization of flow phenomena. *Int. J. Heat Mass Transfer* **16,** 217–219.
Griffin, O. M., and Ramberg, S. E. (1979). Wind tunnel flow visualization with liquid particle aerosols. *In* "Flow Visualization" (T. Asanuma, ed.), 65–73. Hemisphere, Washington, D.C.
Hallock, J. N., Burnham, D. C., Brashears, M. R., Barber, M. R., Tombach, I. H., and Zalay, A. D. (1977). Ground-based measurements of the wake vortex characteristics of a B 747 aircraft in various configurations. *AIAA Pap.* 77-9.
Head, M. R., and Bandyopadhyay, P. (1981). New aspects of turbulent boundary-layer structure. *J. Fluid Mech.* **107,** 297–338.
Hucho, W.-H., and Janssen, L. J. (1979). Flow visualization techniques in vehicle aerodynamics. *In* "Flow Visualization" (T. Asanuma, ed.), pp. 171–180. Hemisphere, Washington, D.C.
Hussain, A. K. M. F., and Clark, A. R. (1981). On the coherent structure of the axisymmetric mixing layer: a flow-visualization study. *J. Fluid Mech.* **104,** 263–294.
Imaizumi, T., Muto, S., and Yoshida, Y. (1979). Flow visualization techniques in an automotive full-scale wind tunnel. *In* "Flow Visualization" (T. Asanuma, ed.), pp. 75–80. Hemisphere, Washington, D.C.
Inaba, H. (1984). Experimental study of natural convection in an inclined air layer. *Int. J. Heat Mass Transfer* **27,** 1127–1139.
Kerker, M., Sculley, M. J., Farone, W. A., and Kassman, A. J. (1978). Optical properties of cigarette smoke aerosols. Appl. Opt. **17,** 3030–3031.
Kimura, S., and Bejan, A. (1983). Mechanism for transition to turbulence in buoyant plume flow. *Int. J. Heat Mass Transfer* **26,** 1515–1532.
Leonard, G. L., Mitchner, M., and Self, S. A. (1983). An experimental study of the electrohydrodynamic flow in electrostatic precipitators. *J. Fluid Mech.* **127,** 123–140.
Linthorst, S. J. M., Schinkel, W. M. M., and Hoogendoorn, C. J. (1981). Flow structure with natural convection in inclined air-filled enclosures. *J. Heat Transfer* **103,** 535–539.
Long, M. B., Chu, B. T., and Chang, R. K. (1980). Instantaneous two-dimensional gas concentration measurements by light scattering. *AIAA Pap.* **80-1370.**
Mach, L. (1896). Über die Sichtbarmachung von Luftstromlinien. *Z. Luftschiffahrt Phys. Atmos.* No. 6, 129–139.
Maltby, R. L., and Keating, R. F. A. (1962). Smoke techniques for use in low speed wind tunnels. *In* "Flow Visualization in Wind Tunnels Using Indicators" (R. L. Maltby, ed.), *AGARDograph* No. 70, 87–109.
McGregor, I. (1961). The vapor-screen method of flow visualization. *J. Fluid Mech.* **11,** 481–511.
McMahon, H. M., Hester, D. D., and Palfery, J. G. (1971). Vortex shedding from a turbulent jet in a cross-wind. *J. Fluid Mech.* **48,** 73–80.
Mensing, P., and Fiedler, H. (1980). Eine Methode zur Sichtbarmachung von hochturbulenten Luftströmungen mit grossen Reynoldszahlen. *Z. Flugwiss. Weltraumforsch.* **4,** 366–369.
Michalke. A. (1964). Zur Instabilität und nichtlinearen Entwicklung einer gestörten Scherschicht. *Ing. Arch.* **33,** 264–276.

Morrison, G. L., and Tran, V. Q. (1978). Laminar flow structure in vertical free convective cavities. *Int. J. Heat Mass Transfer* **21,** 203–213.

Mueller, T. J. (1980). On the historical development of apparatus and techniques for smoke visualization of subsonic and supersonic flow. *AIAA Pap.* **80-0420-CP.**

Mueller, T. J. (1983). Flow visualization by direct injection. *In* "Fluid Mechanics Measurements" (R. J. Goldstein, ed.), pp. 307–375. Hemisphere, Washington, D.C.

Mueller, T. J., and Batill, S. M. (1982). Experimental study of separation on a two-dimensional airfoil at low Reynolds numbers. *AIAA J.* **20,** 457–463.

Mueller, T. J., and Goddard, V. P. (1979). Smoke visualization of subsonic and supersonic flow. *In* "Flow Visualization" (T. Asanuma, ed.), pp. 87–92. Hemisphere, Washington, D.C.

Nagib, H. M. (1979). Visualization of turbulent and complex flows using controlled sheets of smoke streaklines. *In* "Flow Visualization" (T. Asanuma, ed.), pp. 257–263. Hemisphere, Washington, D.C.

Parker, A. G., and Brusse, J. C. (1976). New smoke generator for flow visualization in low speed wind tunnels. *J. Aircr.* **13,** 57–58.

Perry, A. E., Lim, T. T., and Chong, M. S. (1980). The instantaneous velocity fields of coherent structures in coflowing jets and wakes. *J. Fluid Mech.* **101,** 243–256.

Philbert, M., Beaupoil, R., and Faleni, J.-P. (1979). Application d'un dispositif d'éclairage laminaire à la visualisation des écoulements aérodynamiques en soufflerie par émission de fumée. *Rech. Aerosp.* No. 1979-3, 173–179.

Prentice, C. J., and Hurley, F. X. (1970). Subsonic flow visualization using steam. *J. Aircr.* **7,** 380.

Roberts, W. B., and Slovisky, J. A. (1979). Location and magnitude of cascade shock loss by high-speed smoke visualization. *AIAA J.* **17,** 1270–1272.

Ruth, D. W., Hollands, K. G. T., and Raithby, G. D. (1980). On free convection experiments in inclined air layers heated from below. *J. Fluid Mech.* **96,** 461–479.

Sadeh, W. Z., and Brauer, H. J. (1980). A visual investigation of turbulence in stagnation flow about a circular cylinder. *J. Fluid Mech.* **99,** 53–64.

Sakamoto, H., and Arie, M. (1983). Vortex shedding from a rectangular prism and a circular cylinder placed vertically in a turbulent boundary layer. *J. Fluid Mech.* **126,** 147–165.

Sieverding, C. H., and Van den Bosche, P. (1983). The use of coloured smoke to visualize secondary flows in a turbine-blade cascade. *J. Fluid Mech.* **134,** 85–89.

Snow, W. L., and Morris, O. A. (1984). Investigation of light source and scattering medium related to vapor-screen flow visualization in a supersonic wind tunnel. *NASA Tech. Memo.* **NASA TM-86290.**

Sugiyama, S., Hayashi, T., and Yamazaki, K. (1983). Flow characteristics in the curved rectangular channels (Visualization of secondary flow). *Bull. JSME* **26,** 964–969.

Taneda, S. (1978). Visual observations of the flow past a sphere at Reynolds numbers between 10^4 and 10^6. *J. Fluid Mech.* **85,** 187–192.

Torii, K. (1979). Flow visualization by smoke-wire technique. *In* "Flow Visualization" (T. Asanuma, ed.), pp. 251–256. Hemisphere, Washington, D.C.

Véret, C. (1985). Flow visualization by light sheet. *In* "Flow Visualization III" (W.-J. Yang, ed.), pp. 106–112. Hemisphere, Washington, D.C.

Warpinski, N. R., Nagib, H. M., and Lavan, Z. (1972). Experimental investigation of recirculating cells in laminar coaxial jets. *AIAA J.* **10,** 1204–1210.

Wei, Q.-D., and Sato, H. (1984). An experimental study of the mechanism of intermittent separation of a turbulent boundary layer. *J. Fluid Mech.* **143,** 153–172.

Yamada, H. (1973). Instantaneous measurement of air flows by smoke wire technique. *Nippon Kikai Gakkai Ronbunshu* **39,** 726–729.

Yamada, H. (1979). Use of smoke wire technique in measuring velocity profiles of oscillating laminar air flows. *In* "Flow Visualization" (T. Asanuma, ed.), pp. 265–270. Hemisphere, Washington, D.C.

Yu, J. P., Sparrow, E. M., and Eckert, E. R. G. (1972). A smoke generator for use in fluid flow visualization. *Int. J. Heat Mass Transfer* **15,** 557–558.

Zaman, K. B. M. Q., and Hussain, A. K. M. F. (1980). Vortex pairing in a circular jet under controlled excitation. *J. Fluid Mech.* **101,** 449–544.

Zdravkovich, M. M. (1968). Smoke observations of the wake of a group of three cylinders at low Reynolds numbers. *J. Fluid Mech.* **32,** 339–351.

Zdravkovich, M. M. (1973). Smoke visualization of three-dimensional flow patterns in a nominally two-dimensional wake. *J. Mec.* **12,** 225–233.

Section 2.2

Barker, D. B., and Fourney, M. E. (1977). Measuring fluid velocity with speckle patterns. *Opt. Lett.* **1,** 135–137.

Berezin, G. V., and Kudrinskii, V. Z. (1977). Photographic system for determining the motion parameters of solid and liquid particles in a gas stream. *Izv. Vyssh. Uchebn. Zaved., Aviats. Tekh.* **20,** 132–134; *Sov. Aeronaut.* (*Engl. Transl.*) **20,** 104–106.

Bernabeu, E., Amare, J. C., and Arroso, M. P. (1982). White-light speckle method of measurement of flow velocity distribution. *Appl. Opt.* **21,** 2583–2586.

Bez, W., and Frohn, A. (1982). The effect of temperature gradient at the visualization of gas flow. *In* "Flow Visualization II" (W. Merzkirch, ed.), pp. 705–709. Hemisphere, Washington, D.C.

Bicen, A. F. (1982). Refraction correction for LDA measurements in flows with curved optical boundaries. *TSI Q.* **8**(2), 10–12.

Brandone, B., and Bernard, P. (1971). Visualisation par analogie hydraulique de l'écoulement dans une grille d'aubes plane mobile. *Rech. Aerosp.* No. 1971-2, 125–128.

Breslin, J. A., and Emrich, R. J. (1967). Precision measurement of parabolic profile for laminar flow of air between parallel plates. *Phys. Fluids* **10,** 2289–2292.

Carey, V. P., and Gebhard, B. (1982). Transport near a vertical ice surface melting in saline water: experiments at low salinities. *J. Fluid Mech.* **117,** 403–423.

Carlson, D. R., Widnall, S. E., and Peeters, M. F. (1982). A flow-visualization study of transition in plane Poiseuille flow. *J. Fluid Mech.* **121,** 487–505.

Chang, T. P. K., Watson, A. T., and Tatterson, G. B. (1985). Image processing of tracer particle motions as applied to mixing and turbulent flow. Part I: The technique. *Chem. Eng. Sci.* **40,** 269–275.

Charwat, A. F. (1977a). Generator of droplet traces for holographic flow visualization in water tunnels. *Rev. Sci. Instrum.* **48,** 1034–1036.

Charwat, A. F. (1977b). Motion of near-neutrally buoyant tracers in vortical flows. *Phys. Fluids* **20,** 1401–1403.

Chen, C. J., and Emrich, R. J. (1963). Investigation of the shock-tube boundary layer by a tracer method. *Phys. Fluids* **6,** 1–9.

Chiou, C. S., and Gordon, R. J. (1976). Vortex inhibition: Velocity profile measurements *AIChE J.* **22,** 947–950.

Clayton, B. R., and Massey, B. S. (1967). Flow visualization in water: A review of techniques. *J. Sci. Instrum.* **44,** 2–11.

Clift, R., Grace, J. R., and Weber, M. E. (1978). "Bubbles, Drops, and Particles." Academic Press, New York.

Colladay, R. S., and Russell, L. M. (1976). Streakline flow visualization of discrete hole film cooling for gas turbine applications. *J. Heat Transfer* **98,** 245–250.

Coutanceau, M., and Bouard, R. (1977). Experimental determination of the main features of the viscous flow in the wake of a circular cylinder in uniform translation. Part 1: Steady flow. *J. Fluid Mech.* **79,** 231–256.

Coutanceau, M., and Thizon, P. (1981). Wall effect on the bubble behaviour in highly viscous liquids. *J. Fluid Mech.* **107,** 339–373.

Cox, R. G., and Mason, S. G. (1971). Suspended particles in fluid flow through tubes. *Annu. Rev. Fluid Mech.* **3,** 291–318.

Crowe, C. T. (1967). Drag coefficient of particles in a rocket nozzle. *AIAA J.* **5,** 1021–1022.

Denk, V., and Stern, R. (1979). Beitrag zur Kenntnis der Bewegungsvorgänge während der Gärung in zylindrokonischen Gärtanks. *Brauwissenschaft* **32,** 254–262.

De Verdiere, A. C. (1979). Mean flow generation by topographic Rossby waves. *J. Fluid Mech.* **94,** 39–64.

Dimotakis, P. E., Debussy, F. D., and Koochesfahani, M. M. (1981). Particle streak velocity field measurements in a two-dimensional mixing layer. *Phys. Fluids* **24,** 995–999.

Donnelly, R. J. (1981). Fluid dynamics. *In* "AIP 50th Anniversary Physics Vade Mecum" (H. L. Anderson, ed.), pp. 182–195. Am. Inst. Phys., New York.

Douglas, H. A., Mason, P. J., and Hinch, E. J. (1972). Motion due to a moving internal heat source. *J. Fluid Mech.* **54,** 469–480.

Dudderar, T. D., and Simpkins, P. G. (1977). Laser speckle photography in a fluid medium. *Nature (London)* **270,** 45–47.

Durst, F., Melling, A., and Whitelaw, J. H. (1981). "Principles and Practice of Laser-Doppler Anemometry," 2nd Ed. Academic Press, London.

Elkins, R. E., Jackman, G. R., Johnson, R. R., Yoo, J. K., and Lindgren, E. R. (1977). Evaluation of stereoscopic trace particle records of turbulent flow fields. *Rev. Sci. Instrum.* **48,** 738–746.

Emrich, R. J. (1983). Flow field measurement by tracer photography. *Exp. Fluids* **1,** 179–184.

Erbeck, R. (1985). Fast image processing with a microcomputer applied to speckle photography. *Appl. Opt.* **24,** 3838–3841.

Ewan, B. C. R. (1979). Holographic particle velocity measurement in the Fraunhofer plane. *Appl. Opt.* **18,** 623–626.

Fage, A., and Townend, H. C. H. (1932). An examination of turbulent flow with an ultramicroscope. *Proc. R. Soc. London, Ser. A* **135,** 656–677.

Fercher, A. F., and Briers, J. D. (1981). Flow visualization by means of single-exposure speckle photography. *Opt. Commun.* **37,** 326–330.

Françon, M. (1979). "Laser Speckle and Application in Optics." Academic Press, New York.

Gärtner, U., Wernekinck, U., and Merzkirch, W. (1986). Velocity measurements in the field of an internal gravity wave by means of speckle photography. *Exp. Fluids* **4,** 283–287.

Gaster, M. (1964). A new technique for the measurement of low fluid velocities. *J. Fluid Mech.* **20,** 183–192.

Gau, C., and Viskanta, R. (1983). Flow visualization during solid-liquid phase change heat transfer. I: Freezing in a rectangular cavity. *Int. Commun. Heat Mass Transfer* **10,** 173–181.

Gencelli, O. F., Schemm, J. B., and Vest, C. M. (1980). Measurement of size and concentration of scattering particles by speckle photography. *J. Opt. Soc. Am.* **70,** 1212–1218.

Gent, P. R., and Leach, H. (1976). Baroclinic instability in an eccentric annulus. *J. Fluid Mech.* **77,** 769–788.

Greenway, M. E., and Wood, C. J. (1973). The effect of a bevelled trailing edge on vortex shedding and vibration. *J. Fluid Mech.* **61,** 323–335.

Grousson, R., and Mallick, S. (1977). Study of the distribution of velocities in a fluid by speckle photography. *Proc. Soc. Photo-Opt. Instrum. Eng.* **136,** 266–269.

Halow, J. S., and Wills, G. B. (1970). Radial migration of spherical particles in Couette systems. *AIChE J.* **16,** 281–286.

Haselton, F. R., and Scherer, P. W. (1982). Flow visualization of steady streaming in oscillatory flow through a bifurcating tube. *J. Fluid Mech.* **123,** 315–333.

Hasinger, S. H. (1968). An experiment with particles in a free vortex. *AIAA J.* **6,** 939–940.

Hendriks, F., and Aviram, A. (1982). Use of zinc iodide solutions in flow research. *Rev. Sci. Instrum.* **53,** 75–78.

Hinsch, K., Schipper, W., and Mach, D. (1984). Fringe visibility in speckle velocimetry and the analysis of random flow components. *Appl. Opt.* **23,** 4460–4462.

Hinze, J. O. (1975). "Turbulence," 2nd Ed. McGraw-Hill, New York.

Hjemfelt, A. T., Jr., and Mockros, L. F. (1966). Motion of discrete particles in a turbulent fluid. *Appl. Sci. Res.* **16,** 149–161.

Hornemann, U. (1982). Visualization of propellant gas flow inside a gun barrel. *In* "Flow Visualization II" (W. Merzkirch, ed.), pp. 31–36. Hemisphere, Washington, D.C.

Hoyt, J. W., and Taylor, J. J. (1982). Flow visualization using the "floc" technique. *In* "Flow Visualization II" (W. Merzkirch, ed.), pp. 683–887. Hemisphere, Washington, D.C.

Imaichi, K., and Ohmi, K. (1983). Numerical processing of flow visualization pictures: Measurement of two-dimensional vortex flow. *J. Fluid Mech.* **129,** 283–311.

Iwata, K., Hakoshima, T., and Nagata, R. (1978). Measurement of flow velocity distribution by multiple exposure speckle photography. *Opt. Commun.* **25,** 311–314.

Kao, T. W., and Pao, H. P. (1980). Wake collapse in the thermocline and internal solitary waves. *J. Fluid Mech.* **97,** 115–127.

Kao, Y. S., and Kenning, D. B. R. (1972). Thermocapillary flow near a hemispherical bubble on a heated wall. *J. Fluid Mech.* **53,** 715–735.

Kent, J. C., and Eaton, A. R. (1982). Stereophotography of neutral density He-filled bubbles for 3-D fluid motion studies in an engine cylinder. *Appl. Opt.* **21,** 904–912.

Klein, D. E., Miles, J. B., and Bull, S. R. (1980). Pressure drop measurements and flow visualization surrounding roughness elements. *J. Energy* **4,** 112–119.

Koromilas, C. A., and Telionis, D. P. (1980). Unsteady laminar separation: an experimental study. *J. Fluid Mech.* **97,** 347–384.

Lallement, J. P., Desailly, R., and Froehly, C. (1977). Mesure de vitesse dans un liquide par diffusion cohérente. *Acta Astronaut.* **4,** 343–356.

Lee, Y. J., Fourney, M. E., and Moulton, R. W. (1974). Determination of slip ratios in air-water two-phase critical flow at high quality levels utilizing holographic techniques. *AIChE J.* **20,** 209–219.

Mallinson, G. D., Graham, A. D., and De Vahl Davis, G. (1981). Three-dimensional flow in a closed thermo-syphon. *J. Fluid Mech.* **109,** 259–275.

Matisse, P., and Gorman, M. (1984). Neutrally buoyant anisotropic particles for flow visualization. *Phys. Fluids* **27,** 759–760.

Maxworthy, T. (1979). Experiments on the Weis–Fogh mechanism of lift generating by insects in hovering flight. Part 1. Dynamics of the "flying." *J. Fluid Mech.* **93,** 47–63.

McKay, H. C. (1951). "Three-Dimensional Photography." Jones Press, Minneapolis, Minnesota.

Menzel, R., and Shofner, F. M. (1970). An investigation of Fraunhofer holography for velocimetry applications. *Appl. Opt.* **9,** 2073–2079.

Merzkirch, W., and Bracht, K. (1978). The erosion of dust by a shock wave in air: Initial stage with laminar flow. *Int. J. Multiphase Flow* **4,** 89–95.

Meynart, R. (1980). Equal velocity fringes in a Rayleigh–Bénard flow by a speckle method. *Appl. Opt.* **19,** 1385–1386.

Meynart, R. (1982). Digital image processing for speckle flow velocimetry. *Rev. Sci. Instrum.* **53,** 110–111.

Meynart, R. (1983). Instantaneous velocity field measurements in unsteady gas flow by speckle velocimetry. *Appl. Opt.* **22,** 535–540.

Mezaris, T. M., Telionis, D. P., and Jones, G. S. (1982). Visualization of separating oscillatory laminar flow. *In* "Flow Visualization II" (W. Merzkirch, ed.), pp. 259–263. Hemisphere, Washington, D.C.

Morsi, S. A., and Alexander, A. J. (1972). An investigation of particle trajectories in two-phase flow systems. *J. Fluid Mech.* **55,** 193–208.

Nakatani, N., and Yamada, T. (1982). Measurement of non-stationary flow by the pulse-luminescence method using N_2 pulse-laser and polystyrene microcapsules. *In* "Flow Visualization II" (W. Merzkirch, ed.), pp. 215–219. Hemisphere, Washington, D.C.

Nychas, S. G., Hershey, H. C., and Brodkey, R. S. (1973). A visual study of turbulent shear flow. *J. Fluid Mech.* **61,** 513–540.

Orlov, V. V., and Mikhailova, E. S. (1978). An extension of the stroboscopic flow imaging method—The "three-coordinate" technique. *Fluid Mech.—Sov. Res.* **7**(5), 1–13.

Owen, R. B., and Campbell, C. W. (1980). Laser beam manifold and particle photography system for use of fluid velocity measurement. *Rev. Sci. Instrum.* **51,** 1504–1508.

Ozoe, H., Shibata, T., and Churchill, S. W. (1981). Natural convection in an inclined circular cylindrical annulus heated and cooled on its end plates. *Int. J. Heat Mass Transfer* **24,** 727–737.

Pan, F., and Acrivos, A. (1967). Steady flows in rectangular cavities. *J. Fluid Mech.* **28,** 643–655.

Philbert, M., and Boutier, A. (1972). Méthodes optiques de mesure de vitesses de particules entraînées dans les écoulements. *Rech. Aerosp.* No. 1972-3, 171–184.

Powe, R. E., Warrington, R. O., and Scanlan, J. A. (1980). Natural convective flow between a body and its spherical enclosure. *Int. J. Heat Mass Transfer* **23,** 1337–1350.

Rhee, H. S., Koseff, J. R., and Street, R. L. (1984). Flow visualization of a recirculating flow by rheoscopic liquid and liquid crystal techniques. *Exp. Fluids* **2,** 57–64.

Rudinger, G. (1964). Some properties of shock relaxation in gas flows carrying small particles. *Phys. Fluids* **7,** 658–663.

Saffman, P. G. (1965). The lift on a small sphere in a slow shear flow. *J. Fluid Mech.* **22,** 385–400.

Seki, N., Fukusako, S., and Inaba, H. (1978). Visual observation of natural convective flow in a narrow vertical cavity. *J. Fluid Mech.* **84,** 695–704.

Simpkins, P. G., and Dudderar, T. (1978). Laser speckle measurements of transient Bénard convection. *J. Fluid Mech.* **89,** 665–671.

Somerscales, E. F. C. (1981). Tracer methods. *In* "Fluid Dynamics" (R. J. Emrich, ed.), pp. 1–240. Academic Press, New York.

Sparks, G. W., and Ezekiel, S. (1977). Laser streak velocimetry for two-dimensional flows in gases. *AIAA J.* **15,** 110–113.

Suzuki, M., Hosoi, K., Toyooka, S., and Kawahashi, M. (1983). White-light speckle method for obtaining an equi-velocity map of a whole flow field. *Exp. Fluids* **1,** 79–81.

Telionis, D. P., and Koromilas, C. P. (1978). Flow visualization of transient and oscillatory separating laminar flows. *In* "Nonsteady Flow Dynamics" (D. E. Crow and J. A. Miller, eds.), pp. 21–32. Am. Soc. Mech. Eng., New York.

Trinh, E., Zwern, A., and Wang, T. G. (1982). An experimental study of small-amplitude drop oscillations in immisciple liquid systems. *J. Fluid Mech.* **115,** 453–474.
Trolinger, J. D. (1974). Laser instrumentation for flow field diagnostics. *AGARDograph* **AGARD-AG-**186.
Trolinger, J. D., Belz, R. A., and Farmer, W. M. (1969). Holographic techniques for the study of dynamic particle fields. *Appl. Opt.* **8,** 957–961.
Ushizaka, T., and Asakura, T. (1983). Measurements of flow velocity in a microscopic region using a transmission grating. *Appl. Opt.* **22,** 1870–1878.
Van Meel, D. A., and Vermij, H. (1961). A method for flow visualization and measurement of velocity vectors in three-dimensional flow patterns in water models by using colour photography. *Appl. Sci. Res., Sect. A* **10,** 109–117.
Véret, C. (1985). Flow visualization by light sheet. *In* "Flow Visualization III" (W.-J. Yang, ed.), pp. 106–112. Hemisphere, Washington, D.C.
vom Stein, H. D., and Pfeifer, H. J. (1972). Investigation of the velocity relaxation of micron-sized particles in shock waves using laser radiation. *Appl. Opt.* **11,** 305–307.
Weinert, W., Heber, J., and Bayerer, R. (1980). Laser-Stroboskop-Anemometer (LSA) zur Bestimmung von Strömungsvektorfeldern im Hochgeschwindigkeitsbereich. *Z. Flugwiss. Weltraumforsch.* **4,** 137–142.
Yin, S. H., Powe, R. E., Scanlan, J. A., and Bishop, E. H. (1973). Natural convection flow patterns in spherical annuli. *Int. J. Heat Mass Transfer* **16,** 1785–1795.

Section 2.3

Arunachalam, V., Hummel, R. W., and Smith, J. W. (1972). Flow visualization studies of a turbulent drag reducing solution. *Can. J. Chem. Eng.* **50,** 337–343.
Baker, D. J. (1966). A technique for precise measurement of small fluid velocities. *J. Fluid Mech.* **26,** 573–575.
Baker, D. J. (1967). Shear layers in a rotating fluid. *J. Fluid Mech.* **29,** 165–175.
Bhaga, D., and Weber, M. E. (1981). Bubbles in viscous liquids: shapes, wakes and velocities. *J. Fluid Mech.* **105,** 61–85.
Bippes, H. (1977). Experimente zur Entwicklung der freien Wirbel hinter einem Rechteckflügel. *Acta Mech.* **26,** 223–245.
Bippes, H., and Görtler, H. (1972). Dreidimensionale Störungen in der Grenzschicht an einer konkaven Wand. *Acta Mech.* **14,** 251–267.
Brooke-Benjamin, T., and Barnard, B. J. S. (1964). A study of the motion of a cavity in a rotating liquid. *J. Fluid Mech.* **19,** 193–209.
Brown, G. H. (1971). "Techniques of Chemistry. Vol. III: Photochromism." Wiley, New York.
Buzyna, G., and Veronis, G. (1971). Spin-up of a stratified fluid: theory and experiment. *J. Fluid Mech.* **50,** 579–608.
Clamen, M., and Minton, P. (1977). An experimental investigation of flow in an oscillating pipe. *J. Fluid Mech.* **81,** 421–431.
Clutter, D. W., and Smith, A. M. O. (1961). Flow visualization by electrolysis of water. *Aerosp. Eng.* **20,** 24–27, 74–76.
Criminale, W. O., Jr., and Nowell, R. W. (1965). An extended use of the hydrogen bubble flow visualization method. *AIAA J.* **3,** 1203.
Davies, P. A. (1972). Experiments on Taylor columns in rotating stratified fluids. *J. Fluid Mech.* **54,** 691–717.

Davies, P. A., Acheson, D. J., and Titman, C. W. (1976). A note on the geostrophic flow past a crater in a rotating fluid. *Geophys. Fluid Dyn.* **7,** 119–131.

Davis, W., and Fox, R. W. (1967). An evaluation of the hydrogen bubble technique for the quantitative determination of fluid velocities within clear tubes. *J. Basic. Eng.* **89,** 771–781.

Eichhorn, R. (1961). Flow visualization and velocity measurement in natural convection with the tellurium dye method. *J. Heat Transfer* **83,** 379–381.

Gardner, R. A. (1985). Thymol blue flow visualization and modelling of the urban heat island in a density stratified water channel. *In* "Flow Visualization III" (W. J. Yang, ed.), pp. 558–562. Hemisphere, Washington, D.C.

Geller, E. W. (1955). An electrochemical method of visualizing the boundary layer. *J. Aeronaut. Sci.* **22,** 869–870.

Gerrard, J. H. (1971). An experimental investigation of pulsating turbulent water flow in a tube. *J. Fluid Mech.* **46,** 43–64.

Gershenfeld, N., Frazel, R. E., and Whitehead, J. A. (1981). Rotating flume with uniformly flowing, linear stratified water. *Rev. Sci. Instrum.* **52,** 1556–1559.

Grass, A. J. (1971). Structural features of turbulent flow over smooth and rough boundaries. *J. Fluid Mech.* **50,** 233–255.

Hart, J. E. (1971). Transition to a wavy vortex regime in convective flow between inclined plates. *J. Fluid Mech.* **48,** 265–271.

Hearney, J., and Davis, W. (1974). Further contribution to velocity determination using the hydrogen bubble technique. *In* "Flow—Its Measurement and Control in Science and Industry" (R. B. Dowdell, ed.), pp. 833–842. Instrum. Soc. Am., Pittsburgh, Pennsylvania.

Hendriks, F. (1980). Aerodynamics of ink jet printing. *J. Appl. Photogr. Eng.* **6,** 83–86.

Honji, H. (1975). The starting flow down a step. *J. Fluid Mech.* **69,** 229–240.

Honji, H. (1981). Streaked flow around an oscillating circular cylinder. *J. Fluid Mech.* **107,** 509–520.

Humphrey, J. A., Smith, J. W., Davey, B., and Hummel, R. L. (1974). Light-induced disturbances in photochromic flow visualization. *Chem. Eng. Sci.* **29,** 308–312.

Incropera, F. P., and Yaghoubi, M. A. (1980). Buoyancy driven flows originating from heated cylinders submerged in a finite water layer. *Int. J. Heat Mass Transfer* **23,** 269–278.

Iribarne, A., Frantisak, F., Hummel, R. L., and Smith, J. W. (1972). An experimental study of instabilities and other flow properties of a laminar pipe jet. *AIChE J.* **18,** 689–698.

Iritani, Y., Kasagi, N., and Hirata, M. (1983). Direct velocity measurement in low-speed water flows by double-wire hydrogen bubble technique. *Exp. Fluids* **1,** 111–112.

Johnston, J. P., Halleen, R. M., and Lezius, D. K. (1972). Effects of spanwise rotation on the structure of two-dimensional fully developed turbulent channel flow. *J. Fluid Mech.* **56,** 533–557.

Kato, E., Suita, M., and Kawamata, M. (1982). Visualization of unsteady pipe flows using hydrogen bubble technique. *In* "Flow Visualization II" (W. Merzkirch, ed.), pp. 209–213. Hemisphere, Washington, D.C.

Kim, H. T., Kline, S. J., and Reynolds, W. C. (1971). The production of turbulence near a smooth wall in a turbulent boundary layer. *J. Fluid Mech.* **50,** 133–160.

Kimura, S., and Bejan, A. (1980). Experimental study of natural convection in a horizontal cylinder with different end temperatures. *Int. J. Heat Mass Transfer* **23,** 1117–1126.

Kiya, M., Tamura, H., and Arie, M. (1980). Vortex shedding from a circular cylinder in moderate-Reynolds-number shear flow. *J. Fluid Mech.* **101,** 721–735.

Kline, S. J., Reynolds, W. C., Schraub, F. A., and Runstadler, P. W. (1967). The structure of turbulent boundary layers. *J. Fluid Mech.* **30,** 741–773.

Kondratas, H. M., and Hummel, R. L. (1982). Application of the photochromic tracer technique for flow visualization near the wall region. *In* "Flow Visualization II" (W. Merzkirch, ed.), pp. 387–391. Hemisphere, Washington, D.C.

Lavallee, H. C., and Popovich, A. T. (1974). Fluid flow near roughness elements investigated by photolysis method. *Chem. Eng. Sci.* **29,** 49–59.

Matsui, T., Nagata, H., and Yasuda, H. (1979). Some remarks on hydrogen bubble technique for low speed water flows. *In* "Flow Visualization" (T. Asanuma, ed.), pp. 215–220. Hemisphere, Washington, D.C.

Maxworthy, T. (1977). Topographic effects in rapidly-rotating fluids: Flow over a transverse ridge. *J. Appl. Math. Phys.* **28,** 853–864.

McCartney, M. S. (1975). Inertial Taylor columns on a beta plane. *J. Fluid Mech.* **68,** 71–95.

Mochizuki, S., and Yagi, Y. (1982). Characteristics of vortex shedding in plate arrays. *In* "Flow Visualization II" (W. Merzkirch, ed.), pp. 99–103. Hemisphere, Washington, D.C.

Moll, H.-G. (1978). Die Umströmung einer sich frei bewegenden Kugel in einem rotierenden Fluid. *Forsch. Ingenieurwes.* **44,** 25–33.

Moore, M. J., and Long, R. T. (1971). An experimental investigation of turbulent stratified shearing flow. *J. Fluid Mech.* **49,** 635–655.

Nagata, H., Minami, K., and Murata, Y. (1979). Initial flow past an impulsively started circular cylinder. *Bull. JSME* **22,** 512–520.

Nakatani, N., and Yamada, T. (1985). Measurement of non-stationary flow by the pulse-laser induced luminescence method—Increase of observation rate. *In* "Flow Visualization III" (W. J. Yang, ed.), pp. 82–86. Hemisphere, Washington, D.C.

Nakatani, N., Fujiwara, K., Matsumoto, M., and Yamada, T. (1975). Measurement of flow velocity distributions by pulse luminescence method. *J. Phys. E* **8,** 1042–1046.

Nakatani, N., Matsumoto, M., Ohmi, Y., and Yamada, T. (1977). Turbulence measurement by the pulse luminescence method using a nitrogen pulse laser. *J. Phys. E* **10,** 172–176.

Offen, G. R., and Kline, S. J. (1974). Combined dye-streak and hydrogen-bubble visual observation of a turbulent boundary layer. *J. Fluid Mech.* **62,** 223–239.

Popovich, A. T., and Hummel, R. L. (1967). A new method for nondisturbing turbulent flow measurements very close to a wall. *Chem. Eng. Sci.* **22,** 21–25.

Quraishi, M. S., and Fahidy, T. Z. (1980). A technique for the study of flow patterns in electrolysis. *J. Electrochem. Soc.* **127,** 666–669.

Quraishi, M. S., Koroyannakis, D., and Waters, W. F. (1985). A study of combined forced and natural convection flow using an analytical indicator flow visualization technique. *In* "Flow Visualization III" (W. J. Yang, ed.), pp. 743–747. Hemisphere, Washington, D.C.

Rockwell, D. (1972). External excitation of planar jets. *J. Appl. Mech.* **39,** 883–890.

Rockwell, D., and Kniseley, C. (1979). The organized nature of flow impingement upon a corner. *J. Fluid Mech.* **93,** 413–432.

Roos, F. W., and Willmarth, W. W. (1969). Hydrogen bubble visualization at low Reynolds numbers. *AIAA J.* **7,** 1635–1637.

Schetz, J. A., and Eichhorn, R. (1964). Natural convection with discontinuous wall-temperature variations. *J. Fluid Mech.* **18,** 167–176.

Schraub, F. A., Kline, S. J., Henry, J., Runstadler, P. W., Jr., and Littell, A. (1965). Use of hydrogen bubbles for quantitative determination on time-dependent velocity fields in low-speed water flows. *J. Basic Eng.* **87,** 429–444.

Seeley, L. E., Hummel, R. L., and Smith, J. W. (1975). Experimental velocity profiles in

laminar flow around spheres at intermediate Reynolds numbers. *J. Fluid Mech.* **68,** 591–608.
Smith, C. R., and Metzler, S. P. (1983). The characteristics of low-speed streaks in the near-wall region of a turbulent boundary layer. *J. Fluid Mech.* **129,** 27–54.
Smith, C. R., and Paxson, R. D. (1983). A technique for evaluation of three-dimensional behavior in turbulent boundary layers using computer augmented hydrogen bubble-wire flow visualization. *Exp. Fluids* **1,** 43–49.
Sparrow, E. M., Chrysler, G. M., and Azevedo, L. F. (1984). Observed flow reversals and measured-predicted Nusselt numbers for natural convection in a one-sided heated vertical channel. *J. Heat Transfer* **106,** 325–332.
Taneda, S., Honji, H., and Tatsuno, M. (1979). The electrolytic precipitation method of flow visualization. *In* "Flow Visualization" (T. Asanuma, ed.), pp. 209–214. Hemisphere, Washington, D.C.
Thomas, W. C., and Rice, J. C. (1973). Application of the hydrogen-bubble technique for velocity measurements in thin liquid films. *J. Appl. Mech.* **40,** 321–325.
Tory, A. C., and Haywood, K. H. (1968). Flow visualization in the lecture theatre. *Bull. Mech. Eng. Educ.* **7,** 19–24.
van Steenhoven, A. A., and van Dongen, M. E. H. (1979). Model studies of the closing behaviour of the aortic valve. *J. Fluid Mech.* **90,** 21–32.
Vedhanayagam, M., Lienhard, J. H., and Eichhorn, R. (1979). Method for visualizing high Prandtl number heat convection. *J. Heat Transfer* **101,** 571–573.
Veronis, G., and Yang, C. C. (1972). Nonlinear source-sink flow in a rotating pie-shaped basin. *J. Fluid Mech.* **51,** 513–527.
Wortmann, F. X. (1953). Eine Methode zur Beobachtung und Messung von Wasserströmungen mit Tellur. *Z. Angew. Phys.* **5,** 201–206.
Wortmann, F. X. (1969). Visualization of transition. *J. Fluid Mech.* **38,** 473–480.

Section 2.4

Allen, J. R. L. (1966). Note on the use of plaster of paris in flow visualization, and some geological applications. *J. Fluid Mech.* **25,** 331–335.
Bandettini, A., and Peake, D. J. (1979). Diagnosis of separated flow regions on wind-tunnel models using an infrared camera. *ICIASF '79 Rec.* (*IEEE Publ.* **79CH1500-8AES**), 171–185.
Boylan, D. E., Carver, D. B., Stallings, D. W., and Trimmer, L. L. (1978). Measurement and mapping of aerodynamic heating using a remote infrared scanning camera in continuous flow wind tunnels. *AIAA Pap.* **78-799.**
Cérésuela, R., Bétremieux, A., and Cadars, J. (1965). Mesure de l'échauffement cinétique dans les souffleries hypersoniques au moyen de peintures thermosensibles. *Rech. Aerosp.* No. 109, 13–19.
Cerisier, P., Pantaloni, J., Finiels, G., and Amalric, R. (1982). Thermovision applied to Bénard–Marangoni convection. *Appl. Opt.* **21,** 2153–2159.
Chandrasekhar, S. (1977). "Liquid Crystals." Cambridge Univ. Press, London and New York.
Cooper, T. E., Field, R. J., and Meyer, J. F. (1975). Liquid crystal thermography and its application to the study of convective heat transfer. *J. Heat Transfer* **97,** 442–450.
Creel, T. R., Jr., and Hunt, J. L. (1972). Photographing flow fields and heat transfer patterns in color simultaneously. *Aeronaut. Astronaut.* pp. 54–55.
Crowder, J. P. (1982). Fluorescent minitufts for nonintrusive surface flow visualization. *In*

"Flow Visualization II" (W. Merzkirch, ed.), pp. 663–667. Hemisphere, Washington, D.C.

Drummond, J. P., Jones, R. A., and Ash, R. L. (1976). Effective thermal property improves phase change paint data. *AIAA J.* **14,** 1476–1478.

Emery, J. C., Barber, J. B., and Sterrett, J. R. (1967). Flow visualization of a secondary jet by means of lampblack injection techniques. *AIAA J.* **5,** 1039–1040.

Gauffre, G., and Fontanella, J.-C. (1980). Les caméras infrarouges: principes, caractérisation, utilisation. *Rech. Aerosp.* No. 1980-4, 259–269.

George, A. R. (1981). Aerodynamic effects of shape, camber, pitch, and ground proximity on idealized ground-vehicle bodies. *J. Fluids Eng.* **103,** 631–638.

Gessner, F. B., and Chan, Y. L. (1983). Flow in a rectangular diffuser with local flow detachment in the corner region. *J. Fluids Eng.* **105,** 204–211.

Goldstein, R. J., and Timmers, J. F. (1982). Visualization of heat transfer from arrays of impinging jets. *Int. J. Heat Mass Transfer* **25,** 1857–1868.

Hippensteele, S. A., Russel, L. M., and Stepka, F. S. (1983). Evaluation of a method for heat transfer measurements and thermal visualization using a composite of a heater element and liquid crystals. *J. Heat Transfer* **105,** 184–189.

Ishihara, T., Kobayashi, T., and Iwanaga, M. (1982). Visualization of laminar separation by oil film method. *In* "Flow Visualization II" (W. Merzkirch, ed.), pp. 283–287. Hemisphere, Washington, D.C.

Jones, R. A., and Hunt, J. L. (1966). Use of fusible temperature indicators for obtaining quantitative aerodynamic heat transfer data. *NASA Tech. Rep.* **NASA TR R-230.**

Kafka, G., Gaz, J., and Yee, W. T. (1965). Measurement of aerodynamic heating of wind-tunnel models by means of temperature sensitive paint. *J. Spacecr.* **2,** 475–477.

Kang, Y., Nishino, J., Suzuki, K., and Sato, T. (1982). Application of flow and surface temperature visualization techniques to a study of heat transfer in recirculating flow regions. *In* "Flow Visualization II" (W. Merzkirch, ed.), pp. 77–81. Hemisphere, Washington, D.C.

Keener, E. R. (1983). Oil flow separation patterns on an ogive forebody. *AIAA J.* **21,** 550–556.

Kell, D. M. (1978). A surface flow visualization technique for use in cryogenic wind tunnels. *Aeronaut. J.* pp. 484–487.

Keyes, J. W. (1976). Shock interference peak heating measurements using phase change coatings. *J. Spacecr. Rockets* **13,** 61–63.

Kitamura, K., Koike, M., Fukuoka, I., and Saito, T. (1985). Large eddy structure and heat transfer of turbulent natural convection along a vertical flat plate. *Int. J. Heat Mass Transfer* **28,** 837–850.

Klein, E. J. (1968a). Application of liquid crystals to boundary-layer visualization. *AIAA Pap.* **68-376.**

Klein, E. J. (1968b). Liquid crystals in aerodynamic testing. *Astronaut. Aeronaut.* pp. 70–83.

Korkegi, R. H. (1976). On the structure of three-dimensional shock-induced separated flow regions. *AIAA J.* **14,** 597–600.

Kottke, V. (1982). A chemical method for flow visualization and determination of local mass transfer. *In* "Flow Visualization II" (W. Merzkirch, ed.), pp. 657–661. Hemisphere, Washington, D.C.

Kottke, V., Blenke, H., and Schmidt, H. G. (1977). Eine remissionsfotometrische Messmethode zur Bestimmung örtlicher Stoffübergangskoeffizienten bei Zwangskonvektion in Luft. *Waerme- Stoffuebertrag.* **10,** 9–21.

Langston, L. S., and Boyle, M. T. (1982). A new surface-streamline flow-visualization technique. *J. Fluid Mech.* **125,** 53–57.

Maegley, W. J., and Carroll, H. R. (1982). MX missile thermal mapping and surface flow results. *J. Spacecr. Rockets* **19,** 199–204.

Maltby, R. L., ed. (1962). Flow visualization in wind tunnels using indicators. *AGARDograph* **AGARD-AG-**70.

Maltby, R. L., and Keating, R. F. A. (1962). The surface oil flow technique for use in low speed wind tunnels. *AGARDograph* **AGARD-AG-70,** 29–38.

Marvin, J. G., Seegmiller, H. L., Lockman, W. K., Mateer, G. G., Pappas, C. C., and De Rose, C. E. (1972). Surface flow patterns and aerodynamic heating on space shuttle vehicles. *J. Spacecr.* **9,** 573–579.

Meznarsic, V. F., and Gross, L. W. (1982). Experimental investigation of a wing with controlled midspan flow separation. *J. Aircr.* **19,** 435–441.

Monson, D. J. (1983). A nonintrusive laser interferometer method for measurement of skin friction. *Exp. Fluids* **1,** 15–22.

Murai, H., Ihara, A., and Narasaka, T. (1982). Visual investigation of formation process of oil-flow pattern. *In* "Flow Visualization II" (W. Merzkirch, ed.), pp. 629–633. Hemisphere, Washington, D.C.

Nachtsheim, P. R., and Larson, H. K. (1971). Crosshatched ablation patterns in teflon. *AIAA J.* **9,** 1608–1614.

Ogden, T. R., and Hendricks, E. W. (1984). Liquid crystal thermography in water tunnels. *Exp. Fluids* **2,** 65–66.

Peake, D. J., Rainbird, W. J., and Atraghji, E. G. (1972). Three-dimensional flow separations on aircraft and missiles. *AIAA J.* **10,** 567–580.

Peterson, J. I., and Fitzgerald, R. V. (1980). New technique of surface flow visualization based on oxygen quenching of fluorescence. *Rev. Sci. Instrum.* **51,** 670–671.

Pontézière, J., and Bétremieux, A. (1967). Méthode de mesures des flux thermiques pariétaux par peintures thermosensibles. *Assoc. Tech. Marit. Aeronaut., Sess. 1967, Paris* pp. 1–19.

Reding, J. P., and Ericsson, L. E. (1982). Flow visualization reveals causes of shuttle nonlinear aerodynamics. *J. Aircr.* **19,** 928–933.

Sadeh, W. Z., Brauer, H. J., and Durgin, J. R. (1981). Dry-surface coating method for visualization of separation on a bluff body. *AIAA J.* **19,** 954–956.

Schöler, H. (1978). Application of encapsulated liquid crystals on heat transfer measurements in the fin-body interaction region at hypersonic speed. *AIAA Pap.* **78-777.**

Segletes, J. A. (1975). Errors in aerodynamic heat transfer measurements when using phase change coating techniques. *J. Spacecr.* **12,**124–126.

Settles, G. S., and Teng, H.-Y. (1983). Flow visualization methods for separated three-dimensional shock wave/turbulent boundary-layer interactions. *AIAA J.* **21,** 390–397.

Settles, G. S., Perkins, J. J., and Bogdonoff, S. M. (1980). Investigation of three-dimensional shock/boundary layer interactions at swept compression corners. *AIAA J.* **18,** 779–785.

Simonich, J. C., and Moffat, R. J. (1982). New technique for mapping heat transfer coefficient contours. *Rev. Sci. Instrum.* **53,** 678–683.

Sparrow, E. M., and Comb, J. W. (1983). Effect of interwall spacing and fluid flow inlet conditions on a corrugated-wall heat exchanger. *Int. J. Heat Mass Transfer* **26,** 993–1005.

Sparrow, E. M., Ramsey, J. W., and Mass, E. A. (1979). Effect of finite width on heat transfer and fluid flow about an inclined rectangular plate. *J. Heat Transfer* **101,** 199–204.

Sparrow, E. M., Vemuri, S. B., and Kadle, D. S. (1983). Enhanced and local heat transfer, pressure drop, and flow visualization for arrays of block-like electronic components. *Int. J. Heat Mass Transfer* **26,** 689–699.

Squire, L. C. (1962). The motion of a thin oil sheet under the boundary layer on a body. *AGARDograph* **AGARD-AG-70,** 7–28.

Stanbrook, A. (1962). The surface oil flow technique for use in high speed wind tunnels. *AGARDograph* **AGARD-AG-70,** 39–49.

Swigart, R. J. (1974). Cross-hatching studies—A critical review. *AIAA J.* **12,** 1301–1318.

Szodruch, J., and Monson, D. J. (1982). Messung und Sichtbarmachung der leeseitigen Wandschubspannungen bei Deltaflügeln im Überschall. *Z. Flugwiss. Weltraumforsch.* **6,** 279–283.

Tanner, L. H. (1977). Two accurate optical methods for Newtonian viscosity measurement, and observations on a surface effect with silicone oil. *J. Phys. E* **10,** 1019–1028.

Tanner, L. H. (1979). Skin friction measurement by viscosity balance in air and water flows. *J. Phys. E* **12,** 610–619.

Tanner, L. H. (1982). Surface flow visualization and measurement by oil film interferometry. *In* "Flow Visualization II" (W. Merzkirch, ed.), pp. 613–617. Hemisphere, Washington, D.C.

Tanner, L. H., and Blows, L. G. (1976). A study of the motion of oil films on surfaces in air flow, with application to the measurement of skin friction. *J. Phys. E* **9,** 194–202.

Tien, K. K., and Sparrow, E. M. (1979). Local heat transfer and fluid flow characteristics for airflow oblique or normal to a square plate. *Int. J. Heat Mass Transfer* **22,** 349–360.

3

Optical Flow Visualization

3.1. Fluid Flow as a Refractive Index Field

The methods we deal with in this section are essentially those providing the desired information from a light beam, which is transmitted through the flow field (see Section 1.2). The term "optical" is used here, because the information to be recorded, the visible pattern, depends on the variation of an optical property of the fluid, the refractive index. In general, variations of the refractive index remain invisible to the naked eye, and it requires certain optical methods for making such changes visible. Except for one case (streaming birefringence), the variation of the refractive index is caused by changes of the fluid density. Examining when this occurs, we can list the following classes of flows to which optical flow visualization may be applied:

(1) *Compressible flow.* Occurring mostly in gases as high-speed flows; the most dramatic change of density is a shock wave.

(2) *Convective heat transfer.* Changing density, of both gases and liquids, caused by variations of the fluid temperature. Assuming a constant pressure, particularly done for natural or free convection, results in a simple proportionality between the relative changes of density and temperature.

(3) *Mixing.* Mixing of two or more fluids of different density. The density of the mixture varies in the mixing zone according to the local concentration of the individual components.

(4) *Combustion.* Involving all three former effects: high-temperature differences, mixing of several components, and compressibility.

(5) *Plasma flow.* Occurring at very high temperatures when an electron gas (free electrons) may play an important role besides atomic and molecular gases.
(6) *Stratified flow.* Occurring in both gases and liquids; the undisturbed fluid exhibits a vertical density profile or stratification, which can be altered by any motion or flow occurring in this fluid.
(7) *Streaming birefringence.* Only case when the refractive index change is not associated with a density variation. Instead, the refractive index change is caused by the fluid becoming birefringent according to the local distribution of internal shear.

A light beam passing through the test field is disturbed owing to the inhomogeneous distribution of the refractive index in the flowing fluid. The simultaneously occurring alterations with respect to the undisturbed case without flow are of two types: (1) the light is deflected from its original direction and (2) the phase of the disturbed light wave is shifted with respect to that of the undisturbed. Both phenomena can be used not only to visualize the flow but also to perform quantitative density measurements. It is evident that an important requirement for quantitative testing is the exact knowledge of the relation between refractive index and fluid density, which will be discussed in the following.

Since the days of their early development, the methods of optical flow visualization and their application to the mentioned fields of research have continuously been reviewed in books and comprehensive articles [see, e.g., Barnes (1953), Weinberg (1963), Hauf and Grigull (1970), Witte and Collins (1971), Goldstein (1976, 1983), Merzkirch (1981), and Lauterborn and Vogel (1984)].

3.1.1. Relation between Fluid Density and Refractive Index

We consider a light beam of finite width that is transmitted through a flow with varying fluid density. The molecules of the fluid are assumed to have no net electric dipole moment. The light is an electromagnetic wave with the electric field vector $\mathbf{E}$ that distorts the charge configuration of the fluid molecules. A dipole moment $\mathbf{p}$ thereby induced per molecule is proportional to $\mathbf{E}$:

$$\mathbf{p} = \alpha \mathbf{E}, \tag{3.1}$$

where α is the electronic polarizability. Since $\mathbf{E}$ describes an oscillating field, the distortion of the electronic charge configuration is frequency dependent. The field $\mathbf{E}$ might be represented by

$$\mathbf{E} = \mathbf{E}_0 \exp(i2\pi\nu t),$$

where $\mathbf{E}_0$ is the amplitude; ν, the (single) frequency; and t, the time coordinate. In the classical radiation interaction theory of Lorentz, $\mathbf{p}$ is related to $\mathbf{E}$ by utilizing the model of an induced harmonic electron oscillator. If one assumes furthermore that several distorted electrons per molecule with different resonant frequencies ν_i and oscillator strengths f_i contribute to the induced dipole moment, one obtains

$$\mathbf{p} = \frac{e^2\mathbf{E}}{4\pi^2 m_e} \sum_i \frac{f_i}{\nu_i^2 - \nu^2} \tag{3.2}$$

where e is the charge and m_e the mass of an electron. The use of Eq. (3.2) requires us to make the assumption that the resonant frequencies ν_i of the fluid are far from the frequency ν of the light used.

In the preceding discussion it has been presumed that the induced dipole per molecule depends on the external field vector $\mathbf{E}$ only. The fluid molecules in the neighborhood of the molecule under consideration also become electric dipoles and, therefore, generate a secondary electric field, which is superimposed on the external field $\mathbf{E}$. The influence of this secondary field may be neglected if the average distances between the fluid molecules are large (e.g., in a gas). If this approximation cannot be made one should use, instead of Eq. (3.1), a relation $\mathbf{p} = \alpha\mathbf{E}_{\text{eff}}$, where $\mathbf{E}_{\text{eff}}$ is an effective field vector, which also takes into account the secondary field. For a dielectric medium with randomly distributed molecules, $\mathbf{E}_{\text{eff}}$ is known to be

$$\mathbf{E}_{\text{eff}} = \mathbf{E} + \tfrac{4}{3}\pi\mathbf{p},$$

where $\mathbf{p}$ designates the polarisation vector, that is, the net dipole moment per unit volume of the dielectric medium. With N being the number density of molecules in the fluid, the polarization $\mathbf{p}$ is, according to the Lorentz theory, given by

$$\mathbf{p} = \alpha N(\mathbf{E} + \tfrac{4}{3}\pi\mathbf{p}). \tag{3.3}$$

By using the relation $\mathbf{p} = (\varepsilon - 1)\mathbf{E}/4\pi$, where ε is the dielectric constant of the medium, one may eliminate $\mathbf{p}$ and $\mathbf{E}$ from Eq. (3.3) and express the product $(N\alpha)$ as a function of ε. The number density N can be replaced by the density ρ of the fluid through $N = \mathcal{L}\rho/M$, $\mathcal{L}$ being Loschmidt's number and M, the molar weight of the fluid. For the refractive index n of the dielectric fluid, one has $n = \sqrt{\varepsilon}$, so that one derives from Eq. (3.3), with the aid of Eqs. (3.1) and (3.2), the following expression:

$$\frac{n^2 - 1}{n^2 + 2} = \frac{\rho\mathcal{L}e^2}{3\pi m_e M} \sum_i \frac{f_i}{\nu_i^2 - \nu^2}. \tag{3.4}$$

This is a relation between the refractive index n of a fluid and the fluid density ρ expressed in terms of molecular constants and properties of the fluid, and the frequency ν of the light used. Equation (3.4) is often called the Clausius–Mosotti relation and it applies to both gases and liquids.

Since the refractive index of most gases is very close to one, the Clausius–Mosotti relation can be simplified for the case of a gas by setting $(n^2 - 1) \simeq 2(n - 1)$ and $(n^2 + 2) \simeq 3$. Then

$$n - 1 = \rho \frac{\mathcal{L} e^2}{2\pi m_e M} \sum \frac{f_i}{\nu_i^2 - \nu^2}, \tag{3.5}$$

which is called the Gladstone–Dale relation. The abbreviated form

$$n - 1 = K\rho \tag{3.6}$$

defines the Gladstone–Dale constant K, which has the dimension of $1/\rho$ and depends on certain characteristics of the gas as well as on the frequency or wavelength of the light used. Away from the immediate neighborhood of the resonant frequency ν_i, K is only weakly dispersive, which can be seen if one replaces ν_i by the resonant wavelength λ_i; then

$$K = \frac{e^2 \mathcal{L}}{2\pi c^2 m_e M} \sum \frac{f_i \lambda_i^2 \lambda^2}{\lambda^2 - \lambda_i^2}, \tag{3.7}$$

where c is the light velocity in vacuum. Since usually $\lambda_i^2 \ll \lambda^2$, the sum in Eq. (3.7) does not depend on λ^2 within a first-order approximation.

The weak dependency on wavelength of K is demonstrated for air in Table 3.1. Values of refractive indices or Gladstone–Dale constants for a number of pure gases can be found in many textbooks and handbooks on physical chemistry. These values are either estimated from the molecular

TABLE 3.1

Gladstone–Dale Constant for Air at $T = 288$ K

K(cm^3/g)	Wavelength (μm)
0.2239	0.9125
0.2250	0.7034
0.2259	0.6074
0.2274	0.5097
0.2304	0.4079
0.2330	0.3562

TABLE 3.2

Gladstone–Dale Constants for Different Gases

Gas	K (cm^3/g)	Wavelength (μm)	Temperature (K)
He	0.196	0.633	295
Ne	0.075	0.633	295
Ar	0.157	0.633	295
Kr	0.115	0.633	295
Xe	0.119	0.633	295
H_2	1.55	0.633	273
O_2	0.190	0.589	273
N_2	0.238	0.589	273
CO_2	0.229	0.589	273
NO	0.221	0.633	295
H_2O	0.310	0.633	273
CF_4	0.122	0.633	302
CH_4	0.617	0.633	295
SF_6	0.113	0.633	295

data or determined experimentally. Values of K for gases not commonly used for gas dynamic experiments can be found in some special reports or dissertations (e.g., Lensch, 1977; Burner and Goad, 1980; Stirnberg, 1982). Table 3.2 lists the Gladstone–Dale constants for a number of gases; the value for pure water vapor (H_2O) is of interest for the optical investigation of flames. The values of wavelength and temperature in Table 3.2 indicate under which conditions the K values have been determined.

From the preceding paragraph it follows that a gas is described by a linear relationship between **E** and **p**. Due to this linearity, the refractive index of a gas mixture composed of several constituents is given by

$$n - 1 = \sum K_i \rho_i,$$

where the K_i are the Gladstone–Dale constants and ρ_i, the partial densities of the individual components. If one defines a Gladstone–Dale constant K of the mixture by $n - 1 = K\rho$, ρ being the density of the mixture, one has

$$K = \sum K_i \frac{\rho_i}{\rho} \tag{3.8}$$

where (ρ_i/ρ) designates the mass fractions of the components in the gas mixture. Within a first-order approximation, the Gladstone–Dale constant of air can be determined from Eq. (3.8) by using the respective values of O_2 and N_2. Second-order constituents (noble gases, humidity, pollutants) are accounted for in a number of correlation formulas for the K value or specific refractivity of air (e.g., Jones, 1980). Such a precision usually is beyond what is needed for flow experiments.

A dissociated gas can be considered as a mixture of neutral molecules, atoms, and (eventually) molecular residues, if the neutral gas molecule consists of more than two atoms. In the case of a diatomic gas (e.g., pure oxygen), the coefficient of dissociation α_D is the mass fraction of the atoms in the gas mixture. With all the previous assumptions and neglecting the effects of excited states in the mixture, the Gladstone–Dale relation for the dissociated biatomic gas is

$$n - 1 = \rho\{K_M(1 - \alpha_D) + K_A\alpha_D\}, \tag{3.9}$$

where K_M and K_A are the Gladstone–Dale constants of the neutral or molecule gas and of the atom gas, respectively. The Gladstone–Dale constant of the mixture is a function of temperature, since α_D is known to depend on the temperature. Values of K_A for oxygen and nitrogen have been determined by Alpher and White (1959a), Anderson (1969), and Wettlaufer and Glass (1972) and are found to be independent of the values of α_D. The ratio K_M/K_A for oxygen is nearly one at intermediate wavelengths. In performing interferometric studies of air flows, one may therefore disregard the effect of dissociation up to at least $5000K$ and use the Gladstone–Dale constant of neutral air; for oxygen has a large degree of dissociation at these temperatures, but its value of K_M/K_A is nearly one; whereas, nitrogen, although with a ratio K_M/K_A much different from one, has an almost negligible degree of dissociation.

The same procedure can be applied to determine the Gladstone–Dale constant of an ionized gas. If we consider a monatomic gas, the ionized gas mixture consists of neutral atoms, ions, and free electrons. The degree of ionization α_I is the mass fraction of ions in the mixture, and one may neglect the mass fraction of the electrons. The Gladstone–Dale relation of the ionized gas is therefore written

$$n - 1 = \rho\{(1 - \alpha_I)K_A + \alpha_I K_I\} + N_e K_e', \tag{3.10}$$

where K_A and K_I are the Gladstone–Dale constants of the atom gas and of the ion gas, respectively; N_e is the electron number density (i.e., the number of the electrons per unit volume); and the Gladstone–Dale constant of the electron gas, K_e', has dimensions different from those of K_A and K_I. According to Alpher and White (1959b) one has for argon $K_I = \frac{2}{3}K_A$. From plasma theory it follows for the refractive index n_e of the electron gas

$$n_e - 1 = (\lambda^2 N_e e^2)/(2\pi m_e c^2), \tag{3.11}$$

with e being the charge and m_e the mass of an electron; c, the velocity of light in vacuum; and λ, the wavelength of the light, which is transmitted through the gas mixture. If one measures λ in centimeters, one may also write

$$n_e - 1 = -4.46 \times 10^{-14}\lambda^2 N_e,$$

and it becomes evident that the Gladstone–Dale constant of the electron gas, K_e', is negative and strongly dispersive.

If one compares, at a given wavelength, the Gladstone–Dale relation for the electron gas with that of the neutral atom gas, one realizes that the specific refractivity $(n - 1)$ per electron is greater than the specific refractivity per atom by more than one order of magnitude. The optical behavior of an ionized gas is therefore dominated by the presence of free electrons, even at a low ionization level. Since the Gladstone–Dale constant of the electron gas is negative, the optical response to density changes (e.g., in the form of the fringe shift in an interferogram), is opposite to the response recorded in the neutral gas. This is of some importance in visualizing a strong shock wave, where the production of free electrons and the increase in neutral density cause fringe shifts of opposite direction, in the worst case actually canceling one another.

The preceding analysis of the refractive behavior of gases does not account for the range of frequencies ν near an absorption frequency ν_i of the gas. In this case the refractive index n becomes complex, and the imaginary part of n describes the absorption of the transmitted light in the gas. The exact Gladstone–Dale relation in this frequency range must be found from the quantum mechanical solution for the refractive index (Bershader, 1971). Close to a resonant wavelength, the gas exhibits anomalous dispersive behavior with $(n - 1)$ being many orders of magnitude greater than in the nonresonant frequency regime (Fig. 3.1). Since, in

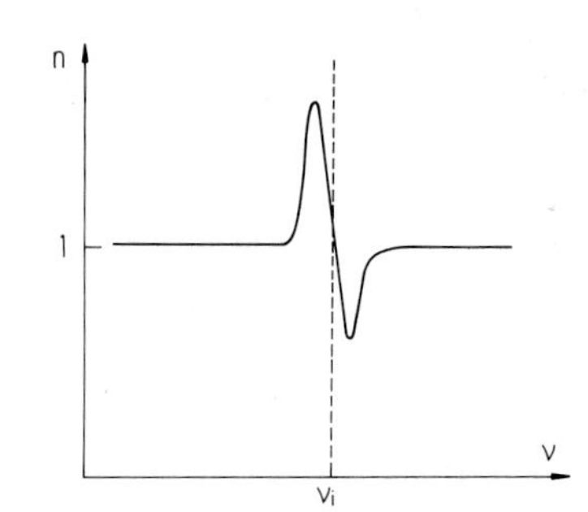

Fig. 3.1 Refractive index of a gas, n, with anomalous dispersion close to a resonant frequency of the gas, ν_i.

addition, the value of the Gladstone–Dale constant is much greater in the resonant regime, the same absolute density change $\Delta\rho$ here will produce much stronger alterations in the refractive index of the gas, Δn, than in the nonresonant case. The possibility of such refractive enhancement has been demonstrated by Blendstrup *et al.* (1978) and by Kügler and Bershader (1983). At a sufficiently high temperature, the test gas is seeded with small quantities of sodium vapor, and a tunable narrow band dye laser, operated near the sodium D_2 line, is used as the light source for interferometric measurements. The sensitivity enhancement is limited by the simultaneously occurring resonance absorption, and the reported sensitivity ratio of 8.3×10^5 (ratio of the specific refractivity of sodium near resonance and that of air under normal conditions; Kügler and Bershader, 1983) is certainly not what can be realized in a simple flow visualization experiment.

The difference $(n - 1)$, also called the refractivity, is by several orders of magnitude larger for liquids than for gases. Therefore, even extremely small density differences in a liquid can produce considerably large signals in the optical methods discussed in the following sections. Fiedler *et al.* (1985) have shown that water being at rest in a normal tank may develop a small difference in temperature between the bottom and the upper surface, say a fraction of a degree, and that the low thermal stratification is sufficient for visualizing flows in such tank with a schlieren system. The relationship between refractive index and density of water (or any other liquid) is hardly available from the Clausius–Mosotti equation, Eq. (3.4), because of the lack of the respective molecular data. The relationship must be determined, therefore, by calibration. Dobbins and Peck (1973) performed a very accurate measurement of the refractive index of water as a function of its temperature in the range between 20 and 34°C. A cubic formula fitting the results is

$$\Delta n_w(T) \times 10^5 = -8.376(T - 20°\text{C}) - 0.2644(T - 20°\text{C})^2 + 0.00479(T - 20°\text{C})^3 \qquad (3.12)$$

where Δn_w is the deviation of the refractive index from the value $n_w = 1.332156$ for $T = 20°\text{C}$, determined for a wavelength of $\lambda = 0.6328$ μm. Note that the temperature is given here in degrees celcius.

Many experiments on stratified flows are performed in saltwater systems. For a quantitative evaluation it is necessary to know the relationship between the refractive index of saltwater and the salinity of the solution. Grange *et al.* (1976) have measured the refractive index n of H_2O–NaCl solutions at $T = 25°\text{C}$ and $\lambda = 0.6328$ μm; Peters (1985) gives n as a function of the density ρ of the H_2O–NaCl solution at $T = 20°\text{C}$ (Fig. 3.2). In both cases the relationships are nearly linear.

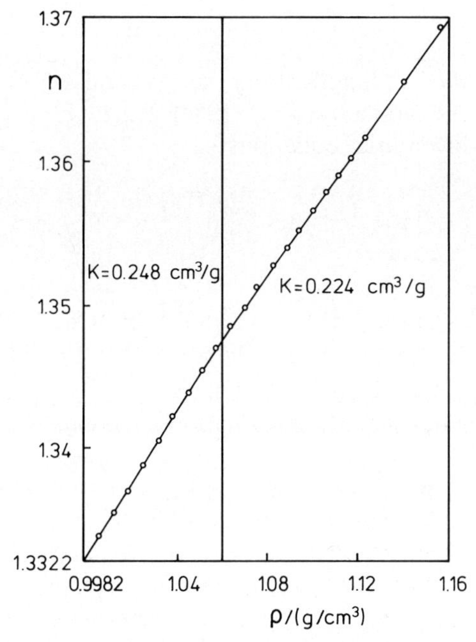

Fig. 3.2 Refractive index of saltwater as function of density measured for a wavelength $\lambda = 632.8$ nm and at a temperature of T = 20°C. In the two sections separated by the vertical solid line, the experimental points can be approximated by linear relationships n = const + $K\rho$ with different values of K. (From Peters, 1985.)

3.1.2. Deflection and Retardation of a Light Ray in an Inhomogeneous Density Field

Having established the relationship between the refractive index and the fluid density, we may now investigate the problem of how a light ray is disturbed in an inhomogeneous refractive field. The problem is treated here in terms of geometrical optics, in accordance to the presentation of Weyl (1954). This means that some physical phenomena, like diffraction or dispersion, are excluded from the analysis. The refractive index n is assumed to vary as a function of the three spatial coordinates x, y, and z in the flow field, i.e., $n = n(x, y, z)$. An incident light beam initially parallel to the z direction is transmitted through the flow. In many practical arrangements, the flow facility has plane viewing windows, which are normal to the z direction. The propagation of a single light ray in the refractive field (also called a phase object) is described by Fermat's principle, which states that the variation of optical path length along a light ray in the object must vanish; hence

$$\delta \int n(x, y, z)\, ds = 0, \tag{3.13}$$

where s denotes the arc length along the ray, and ds is defined by $ds^2 = dx^2 + dy^2 + dz^2$. As shown by Weyl (1954), Eq. (3.13) is equivalent to the following set of differential equations:

$$\frac{d^2x}{dz^2} = \left\{1 + \left(\frac{dx}{dz}\right)^2 + \left(\frac{dy}{dz}\right)^2\right\}\left\{\frac{1}{n}\frac{\partial n}{\partial x} - \frac{dx}{dz}\frac{1}{n}\frac{\partial n}{\partial z}\right\}$$

$$\frac{d^2y}{dz^2} = \left\{1 + \left(\frac{dx}{dz}\right)^2 + \left(\frac{dy}{dz}\right)^2\right\}\left\{\frac{1}{n}\frac{\partial n}{\partial y} - \frac{dy}{dz}\frac{1}{n}\frac{\partial n}{\partial z}\right\} \tag{3.14}$$

The problem is to determine a solution of this system, $x = x(z)$ and $y = y(z)$, which describes the path of the light ray through the refractive index field.

The solution of the system [Eqs. (3.14)] must be found for a certain initial condition, which specifies the particular light ray in the transmitted, parallel beam. This initial condition is given by specifying the coordinates $x = \xi_1$ and $y = \eta_1$, where the ray enters the test volume. Having found the particular solution of Eq. (3.14), one then determines the coordinates ξ_2 and η_2 of the ray in the exit plane of the flow test volume as well as the respective inclinations of the ray at the exit point. In an optical arrangement, one usually has a recording or photographic plane at a distance l from the exit plane of the test volume. In order to predict the observable pattern in the recording plane one may determine, for each ray, the following three quantities (see Fig. 3.3): (1) the displacement **QQ*** of a deflected or disturbed ray with respect to an undisturbed ray, that is, a ray, which has passed through a homogeneous field; (2) the deflection angles ε_x and ε_y of the ray at the end of the test volume; and (3) the

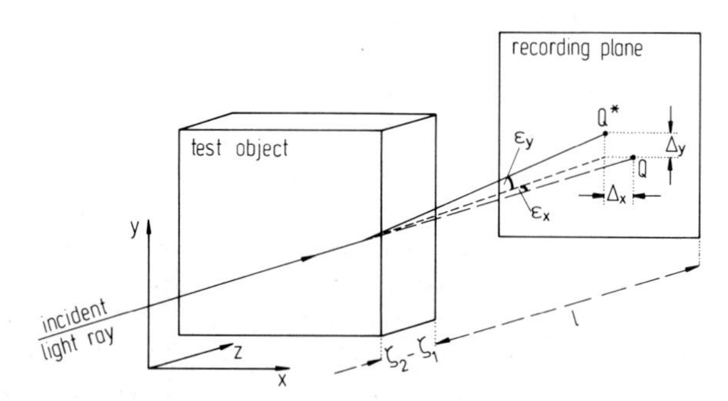

Fig. 3.3 Deflection of a light ray in an inhomogeneous test object (density field): geometric relations.

retardation of the disturbed ray with respect to an undisturbed ray. The latter quantity can be expressed by the time difference Δt between the arrival of the two rays in the recording plane. It will be shown later that each of these three quantities corresponds to a particular class of optical visualization methods.

The form of Eq. (3.14) is too complicated to permit an overall perspective. It is helpful to simplify the system by making the following assumptions: it is a fact of experience that the deviations from the z direction of a light ray in a compressible gas flow are negligibly small, but that the ray may leave the test field with a nonnegligible curvature. Hence, one may assume that the slopes of the ray, dx/dz and dy/dz, are very small everywhere as compared with unity; and since in most cases $\partial n/\partial x$, $\partial n/\partial y$, and $\partial n/\partial z$ are of the same order of magnitude, the simplified system of equations reads

$$\frac{d^2x}{dz^2} = \frac{1}{n}\frac{\partial n}{\partial x}, \qquad \frac{d^2y}{dz^2} = \frac{1}{n}\frac{\partial n}{\partial y}. \tag{3.15}$$

Due to the simplification introduced, the light ray now leaves the test field at the same coordinates ξ, η where it enters the field, but at an angle that is determined by the line integral of the derivative of the refractive index distribution in the test field.

With the aid of the reduced set of equations [Eq. (3.15)], one may determine the three aforementioned observable quantities. Expressing the displacement $\mathbf{QQ}^*$, and the deflection angle ε in terms of the respective x and y components, one obtains (see Fig. 3.3):

$$(\mathbf{QQ}^*)_x = l \int_{\zeta_1}^{\zeta_2} \frac{1}{n}\frac{\partial n}{\partial x}\, dz,$$

$$(\mathbf{QQ}^*)_y = l \int_{\zeta_1}^{\zeta_2} \frac{1}{n}\frac{\partial n}{\partial y}\, dz, \tag{3.16a}$$

$$\tan \varepsilon_x = \int_{\zeta_1}^{\zeta_2} \frac{1}{n}\frac{\partial n}{\partial x}\, dz,$$

$$\tan \varepsilon_y = \int_{\zeta_1}^{\zeta_2} \frac{1}{n}\frac{\partial n}{\partial y}\, dz, \tag{3.16b}$$

$$\Delta t = \frac{1}{c} \int_{\zeta_1}^{\zeta_2} \{n(x, y, z) - n_\infty\}\, dz. \tag{3.16c}$$

In Eq. (3.16c) c is the velocity of light in vacuum, n_∞, the refractive index of the undisturbed test field in which the reference ray propagates; and ζ_1

and ζ_2 are the z coordinates of the points where a ray enters and leaves the test field. The recording plane is a distance l from the exit plane of the test field, and it has been assumed that the gas in the regime between the flow facility and the recording plane has uniform optical behavior. The quantity Δt can be converted into the optical phase difference $\Delta\varphi$ between the disturbed and the undisturbed ray in the recording plane, and one has

$$\frac{\Delta\varphi}{2\pi} = \frac{1}{\lambda}\int_{\zeta_1}^{\zeta_2} \{n(x, y, z) - n_\infty\}\, dz, \tag{3.17}$$

where λ is the wavelength of the light used.

The following sections deal with different optical methods applied to the visualization of compressible flows. The shadowgraph is a technique, which visualizes the displacement $\mathbf{QQ}^*$ as represented by Eq. (3.16a). The schlieren system measures the deflection angle ε described by Eq. (3.16b). Optical phase changes experienced by a light ray in the compressible field according to Eq. (3.17) can be made visible with optical interferometers. As will be shown later, these different classes of visualization method exhibit another systematic behavior: the shadowgraph is sensitive to changes in the second derivative of the gas density ρ, the schlieren system visualizes changes in the first derivative of ρ, and with interferometers one may measure absolute density changes.

The final results, Eqs. (3.16) and (3.17), are derived from the simplified equations [Eq. (3.15)]. There are cases where the simplifying assumption that the individual rays follow straight path lines through the flow cannot be made. The deviation from such a straight path line is caused by a strong density gradient normal to the direction of light propagation, here the z direction. These deviations are particularly strong in liquids due to the high value of the refractivity $(n - 1)$. This effect is demonstrated in Fig. 3.59. A laser beam, initially horizontal, propagates through a glass tank filled with stratified saltwater; i.e., the fluid has a strong density gradient in vertical direction. The laser beam becomes visible due to the sideward scattering of light from particles suspended in the liquid solution. In Section 3.3.6 it is shown how this "strong refraction effect" can be accounted for in the quantitative evaluation of flow interferograms.

3.2. Visualization by Means of Light Deflection

3.2.1. *The Shadowgraph*

In its simplest form, the shadowgraph does not need any optical component, and the effect can therefore be observed in many situations outside of a laboratory. The utilization of the shadow effect for scientific testing and flow visualization has been analyzed first by Dvorak (1880), a co-worker of Ernst Mach. An essential feature of the method is the use of a point-shaped light source. The light diverging from this point source is transmitted through the test field, and the shadow picture produced by the inhomogeneous density field can be recorded in a vertical plane placed at a distance l behind the test field. Instead of this simplest arrangement, a system will be regarded here with parallel light through the flow field, which might be bounded by plane viewing windows normal to the parallel light beam (Fig. 3.4). In order to avoid the use of too great a photographic plate, the recording plane can be focused by means of a camera lens onto a film or plate of reduced size. The contour of a rigid object in the test field appears unfocused on the shadow picture. The sharpness of such objects increases with decreasing diameter of the light source.

When passing through the test field under investigation, the individual light rays are refracted and bent out of their original path. A particular deflected light ray should be traced which arrives at a point Q^* of the recording plane instead of a point Q (Fig. 3.3), so that the distribution of light intensity in that plane is altered with respect to the undisturbed case (e.g., Q^* receives more light than before, while no light arrives at Q, which can be regarded as the shadow of the respective object point). Assume that $I(x, y)$ denotes the light intensity distribution in the recording plane for the undisturbed case, whereas $I^*(x^*, y^*)$ is the intensity for the disturbed case. The intensity I^* at point (x^*, y^*) results from all intensities I_i related to points (x_i, y_i), which are mapped into (x^*, y^*). In determining the resulting intensity I^*, one has to take into account that the area illuminated by a particular light beam is deformed due to the mapping of the (x, y) plane into the (x^*, y^*) plane. The intensity per unit area thereby changes. Each intensity value I_i contributing to the summation has to be divided by a value, which accounts for this mapping, and the denominator of this expression is the Jacobian of the mapping function of the system (x, y) into (x^*, y^*). The intensity resulting in a point (x^*, y^*) is therefore

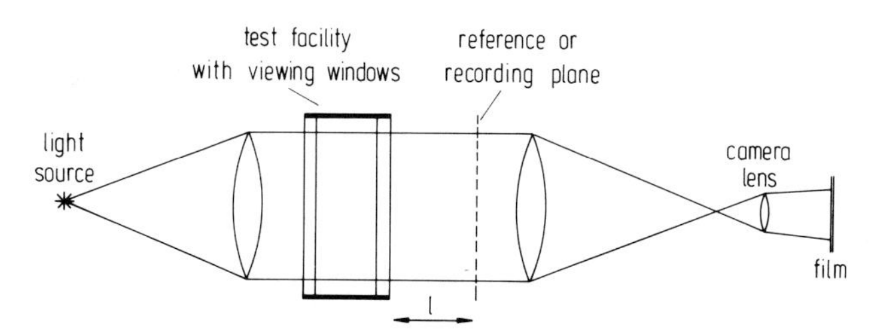

Fig. 3.4 Shadowgraph system with parallel light through the test field.

given by

$$I^*(x^*, y^*) = \sum_i \frac{I_i(x_i, y_i)}{|\partial(x^*, y^*)/\partial(x, y)|}. \tag{3.18}$$

Assume that only one point (x, y) is mapped into (x^*, y^*), and that the new coordinates are given by

$$x^* = x + \Delta x(x, y), \qquad y^* = y + \Delta y(x, y),$$

where Δx and Δy describe the displacement $\mathbf{QQ}^*$ determined by Eq. (3.16a). A linearization of the mapping function is introduced by assuming that Δx and Δy are small quantities, and that products and higher powers of Δx and Δy can be neglected. The Jacobian is then

$$\left|\frac{\partial(x^*, y^*)}{\partial(x, y)}\right| \cong 1 + \frac{\partial \Delta x}{\partial x} + \frac{\partial \Delta y}{\partial y}.$$

A photographic plate is sensitive to relative changes of the light intensity, which is given in this case by $(I^* - I)/I = \Delta I/I$. With the results of the above linearization and Eqs. (3.18) and (3.16a), one obtains

$$\frac{\Delta I}{I} = l \int_{\zeta_1}^{\zeta_2} \left(\frac{\partial^2}{\partial x^2} + \frac{\partial^2}{\partial y^2}\right) (\ln n)\, dz. \tag{3.19}$$

By applying the Gladstone–Dale formula, Eq. (3.6), it becomes evident that the shadowgraph is sensitive to changes in the second derivate of the (gas) density.

The latter result can be visualized by a simple analogy, without using equations. Consider a beam of light traversing a transparent plate of a homogeneous material (e.g., glass) and of constant thickness (Fig. 3.5a).

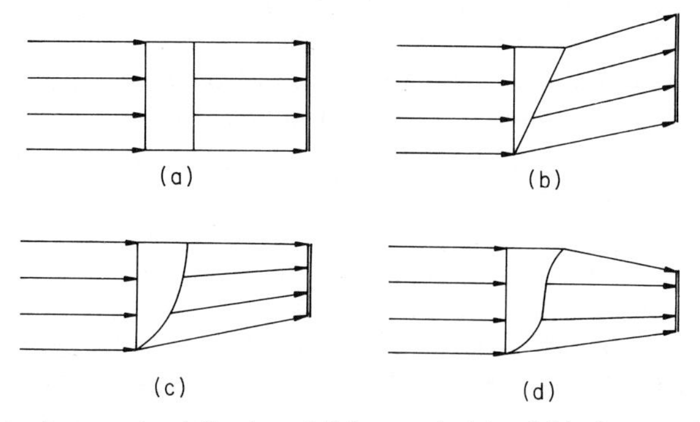

Fig. 3.5 Analogy to the deflection of light rays in (a) a field of constant density, (b) constant density gradient, (c) constant second derivative of density, and (d) variable second derivative of density.

This case is equivalent to a fluid field of constant density. The beam is not affected, and a recording plane behind the object is evenly illuminated. If the beam is directed through a wedge with plane surfaces (Fig. 3.5b), which is equivalent to a fluid field with $\partial n/\partial y$ = const (y being the direction normal to the incident light), then the beam is deflected by a constant angle, and the recording plane again is uniformly illuminated. If one surface of the wedge is of spherical (or cylindrical) shape (Fig. 3.5c), which is equivalent to a fluid field with $\partial^2 n/\partial^2 y$ = const, then it is found that the rays of the beam will be convergent behind the wedge (lens effect), but the illumination of the recording plane, though being brighter, is still uniform. Only if the curvature of the wedge surface is not constant (Fig. 3.5d), which means that $\partial^2 n/\partial^2 y \neq$ const in the fluid, is the illumination of the recording plane nonuniform, and only then is it possible to discriminate between areas of different light intensity and to derive a visible information from the record.

From the result of the brief analysis it follows that the shadowgraph is not a method suitable for quantitative measurement of the fluid density, since such an evaluation would require one to perform a double integration of the data, i.e., the intensity distribution in the recording plane. The intensity, expressed, for example, in shades of gray, cannot be determined very precisely, and the errors would be amplified in the double integration. Furthermore, several simplifications contained in the preceding analysis might be violated. Owing to its simplicity, however, the shadowgraph is a convenient method for obtaining a quick survey of flow fields with varying fluid density.

The most drastic change in the second derivative of n or ρ occurs in a shock wave. The shadowgraph is most appropriate for visualizing the geometry of shock patterns, either a steady configuration in supersonic flow (Fig. 3.6) or unsteady shock configurations as produced in a shock tube (Fig. 3.7). For the measurement of the shock distance from a blunt body in a supersonic flow it is important to know where the exact location of the infinitesimally thin shock surface is, because such a bow shock may appear as a relatively thick band in the shadow picture. Figure 3.8 gives an answer to this question for the case of parallel incident light. The trace of the shock on the screen or photograph is a band of absolute darkness bounded on the downstream side by an edge of intense brightness subsequently shading into normal illumination. The outer edge of the dark zone is the exact geometrical position of the shock front. Owing to the density increase through the shock, the fluid volume behind the shock acts in a way similar to a convex lens. The caustic, which corresponds to the focusing effect of the lens, explains the concentration of light on the downstream side. It also gives an explanation for the observed deforma-

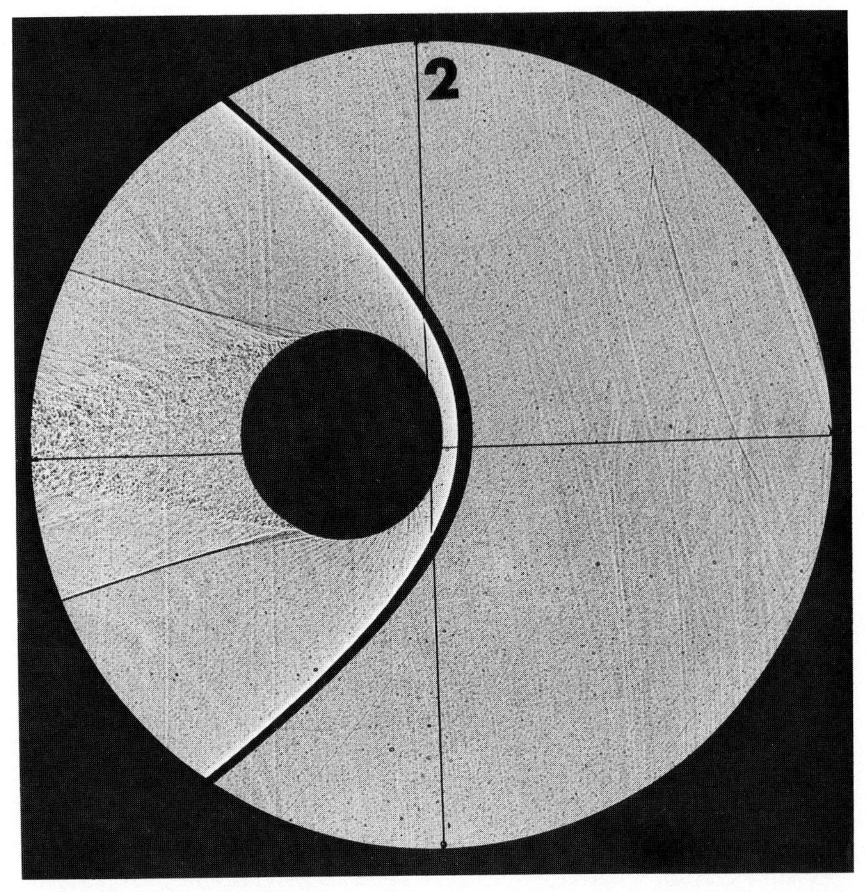

Fig. 3.6 Shadowgraph of a sphere flying at a Mach number $M = 1.7$. (From Stilp, 1968.)

tion of the spherical contour (Fig. 3.6): Light can be deflected onto points of the screen where one should expect to see the shadow of the front part of the sphere. These possible error sources must be taken into account if the aim is to measure the stand-off distance of a bow shock for which the shadowgraph has become a standard method. Since this distance can pulsate during a test run, it is advantageous to record the shock position in real time. Sajben and Crites (1979) and Roos and Bogar (1982) have shown that this can be done by using a camera containing a linear array of photodiodes or pixels. The pixels are sensed by electronic means, and those pixels are identified onto which the shadow of the shock is cast.

Fig. 3.7 Shadowgraph of two-dimensional shock diffraction in a shock tube; vortex formation at the apex of the vertical wedge. (From Schardin, 1958.)

The projection by the sunlight of a shock wave's shadow onto a solid surface can be observed, under certain circumstances, from the window of a jet airplane cruising at high subsonic speed. Then, local supersonic flow regimes occur on the curved upper surface of the wing, and the shadow of the accompanying shock, which brings the air velocity back to a subsonic value, can be traced on the wing as a dark line (Fig. 3.9). This line is seen to oscillate somehow, since the supersonic regime also changes its shape and position during the cruise. Such a projection onto a surface in the flow field is also a means for studying boundary-layer/

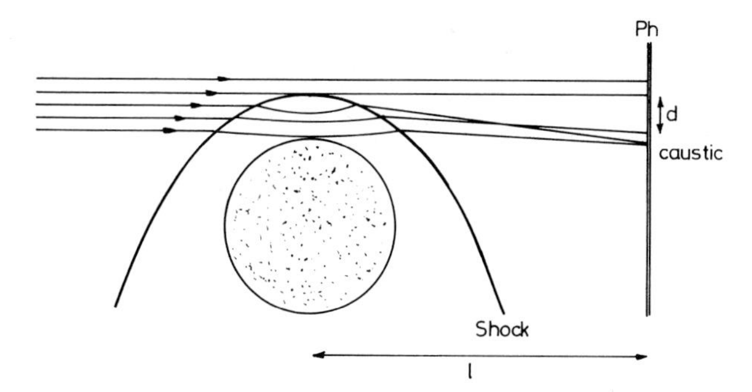

Fig. 3.8 Formation of the shadow of the bow shock wave ahead of a sphere flying at supersonic speed.

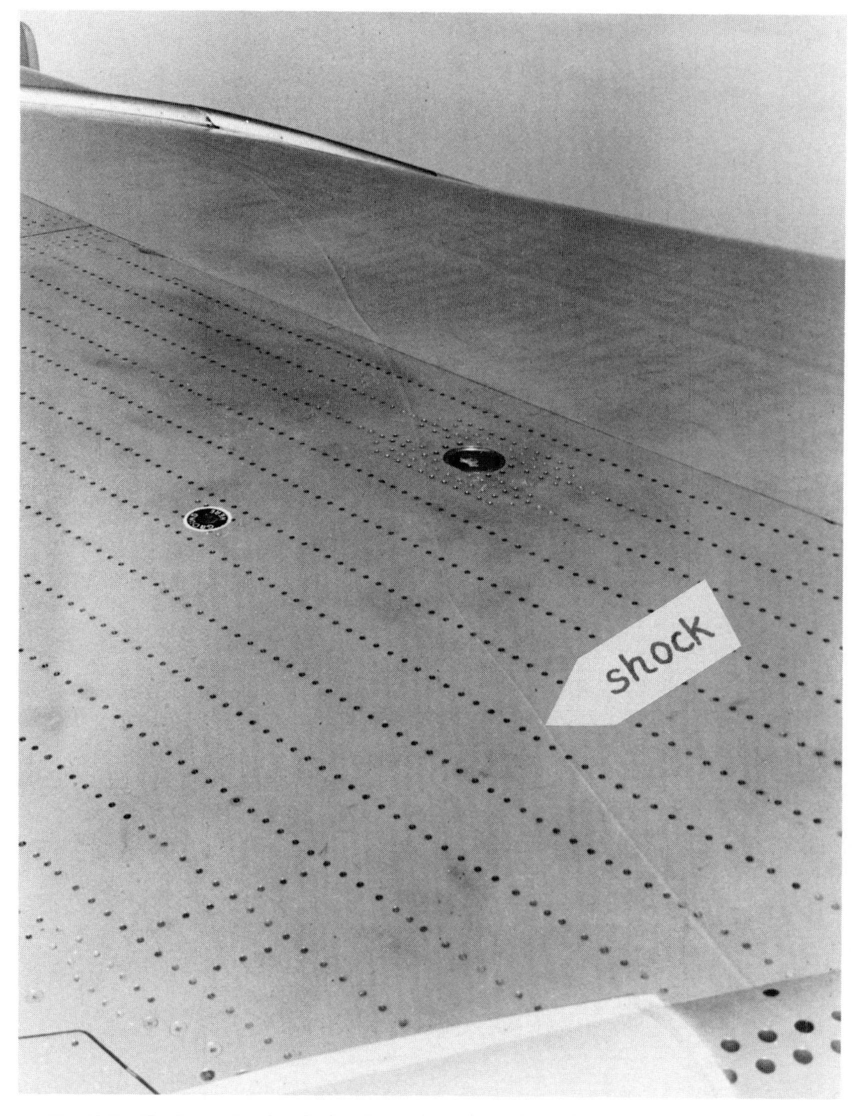

Fig. 3.9 Shadow of a local shock on the wing of an aircraft flying at high subsonic speed. (From Larmore and Hall, 1971.)

shock-wave interactions in a wind tunnel. The projected shadow of the shock together with simultaneously applied oil film visualizations can produce detailed information on the boundary layer flow (Klein, 1970; Sedney, 1972). Disturbances in a uniform, inviscid supersonic flow often display conical rather than planar symmetry. They can be visualized by conical shadowgraphy in which a conical light beam and the conical flow under study have common vertices, and a visual pattern is formed in a plane normal to the cone axis (Schmidt and Settles, 1986).

If the incident light beam is parallel to the plane of a shock wave, and if the shock is regarded as a discontinuity surface, the formation of a shadow of the plane shock could not be described with the laws of geometrical optics. That the shadow image is formed yet in such a case, can be attributed to the following effects: (1) the light is never actually exactly parallel to the discontinuity surface; (2) the shock is of finite thickness; and (3) the major contribution to the formation of the shadow results from light diffraction at the discontinuity surface. Pfeifer *et al.* (1970) have shown that, for a plane shock, the zeroth diffraction order is a minimum, and with white light illumination only this minimum appears in the shadowgraph to form the observed dark band. The same interpretation applies to the formation of the shadow picture of an expansion wave the front of which appears as a bright band, followed by a zone of reduced brightness.

Numerous are the applications of the shadowgraph to problems of compressible flows, mixing of fluids, convective heat transfer, and stratified flows. The observed pattern can be explained to some extent by geometrical ray tracing, e.g., the formation of a bright band, parallel to the wall, in a compressible laminar boundary layer (Fig. 3.10). In a two-

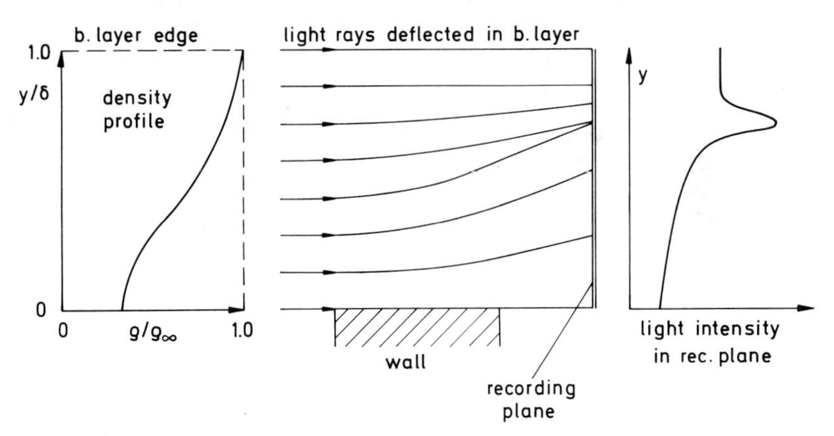

Fig. 3.10 Representation of the density profile in a compressible boundary layer by the shadow method.

dimensional flow, where the fluid parameters do not depend on the coordinate in the direction of the transmitted light, it might be possible to directly measure the amount of light deflection (i.e., the displacement **QQ*** in Fig. 3.3). Bathelt *et al.* (1979) relate the measured displacement to the gradient of the fluid temperature at a solid surface, and thus they can directly deduce the local heat transfer or the Nusselt number from the shadowgraph. A more precise way of determining this quantity is given by the methods that directly react to changes of the first derivative of the refractive index (see the following sections).

Shadowgraphs taken of turbulent flow fields with fluctuating fluid density show a granular structure, which carries information on the turbulent structure (see, e.g., the wake of the sphere in Fig. 3.6). Essential for the generation of such a pattern is an exposure time short enough for "freezing" the turbulent fluctuations. Deduction of quantitative data on the density distribution is rather impossible due to the three-dimensional nature of the turbulent field. However, Uberoi and Kovasznay (1955) already pointed out that it can be possible to statistically analyze the turbulent pattern and to derive certain parameters describing the turbulent field (e.g., characteristic length scales) provided the turbulence in the flow is isotropic. Hesselink and White (1983) tried to quantify this proposal by using digital image processing. But a shortcoming of the method still is the necessity of measuring values of the shades of grey in the shadowgraph and the dependence of the signal on the second derivative of the fluid density. Recent developments show that Uberoi and Kovasznay's idea might be realized better with an optical method yielding discrete number values for the light deflection in the turbulent field (see Section 3.2.4). A single, unexpanded laser beam transmitted through a turbulent flow with fluctuating density produces, due to the shadow effect, a signal, which can be received by an electro-optical device (e.g., a photomultiplier). The signal can be recorded continuously in time and analyzed to provide information on the turbulence (Grandke, 1985). Such a device, however, which also has analogous modifications for the schlieren and the interference effect, is rather an optical probe and not a visualization technique as we have defined in the beginning.

3.2.2. Schlieren Methods

The German word "schliere" designates in a transparent medium a local inhomogeneity, which causes an irregular light deflection. The word "schliere" has been attributed to certain optical devices, that allow one not only to visualize qualitatively such optical inhomogeneities but also to measure quantitatively the amount of deflection. The principle of the method was developed more than a century ago and is attributed, depending on the authors' national preference, to Foucault (1859) and to Toepler (1864). Foucault used this device, of which the principal component is a knife edge (see Fig. 3.11), to check the quality of optical elements like lenses and mirrors, whereas, it was Teopler who recognized and described in his book the wide applicability of this "knife-edge method," including the visualization of compressible flows; the name "Toepler method" has therefore become common in this field.

Today the schlieren method is the most frequently applied optical visualization system in aerodynamic and thermodynamic laboratories, since it combines a relatively simple optical arrangement with a high degree of resolution. Many modifications have been derived from the original system. Review of the various arrangements and discussion of their application, detailed insight into the physical background, and extensive comments on the technical performance of the schlieren methods have been given in the survey articles and in the work of Schardin (1934, 1942), Beams (1954), Wolter (1956), Holder *et al.* (1956), and Holder and North (1963). These articles or books also contain numerous references to papers reporting on the application of the schlieren method for gas dynamic or heat-transfer measurements, as well as in other fields, as indicated in Section 3.1.

3.2.2.1. *Theory of the Toepler Method*

We consider a schlieren system with parallel light through the test section containing a compressible flow field (Fig. 3.11). An image of the light source is formed in the plane of the knife edge, which is placed in the focal plane of the second lens (or spherical mirror), also called the "schlieren head." The knife edge is perpendicular here to the plane of the figure, and the light source is either point-shaped or a narrow slit parallel to the knife edge. The camera lens serves to form an image of the test section on the photographic film, thereby eliminating possible shadow effects.

If the knife edge cuts off a part of the light-source image, the light intensity illuminating the photographic plate will be reduced. Let a be the

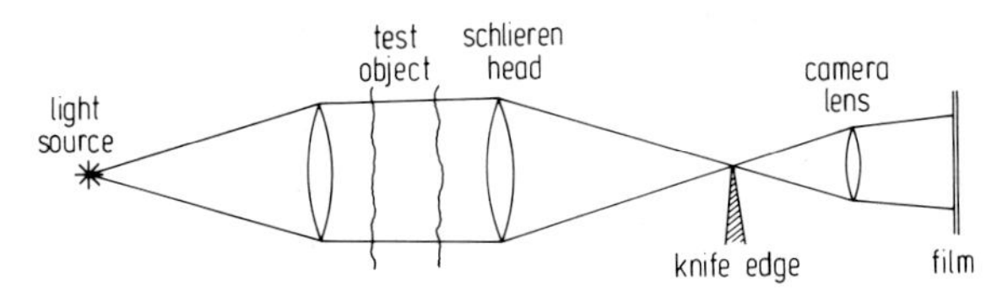

Fig. 3.11 Teopler schlieren system with parallel light through the test field.

reduced height of the light source image and b its width perpendicular to the figure plane; in the absence of any disturbance in the test section, the light intensity I arriving in any point (x, y) of the photographic plane will be constant, and the uniform illumination is given with sufficient accuracy by

$$I(x, y) = \eta I_0(ab/f_c^2) = \text{const}, \tag{3.20}$$

where I_0 is the original intensity (luminance) of the light source, η an absorption coefficient describing the loss of intensity on the way from the light source to the knife-edge plane, and f_c is the focal length of the camera lens in Fig. 3.11. Equation (3.20) is only valid, if geometric-optical aberrations of the system (e.g., coma and astigmatism) can be neglected.

If a number of light rays is deflected by an angle ε due to a disturbance in the test section, the corresponding image of the light source formed by these rays in the plane of the knife edge is shifted by distances Δa and Δb perpendicular and parallel to the edge (Fig. 3.12). If ε_y is the vertical component of ε, and f_2 the focal length of the schlieren head, then

$$\Delta a = f_2 \tan \varepsilon_y \cong \varepsilon_y f_2. \tag{3.21}$$

With parallel light through the test section, Δa is independent of the distance between the test field and the schlieren head. The light intensity in (x, y) is then changed by

$$\Delta I = \eta I_0(\Delta a b/f_c^2). \tag{3.22}$$

We assume thereby, that the absorption coefficient is the same as in the undisturbed case. The photographic process allows one to measure relative intensity changes rather than absolute values; therefore

$$\Delta I/I = \Delta a/a = \varepsilon_y(f_2/a). \tag{3.23}$$

Replacing ε_y (or $\tan \varepsilon_y$) by Eq. (3.16b), we recognize that the local change of light intensity $(\Delta I/I)$ in a schlieren image can be evaluated to obtain the gradient of the refractive index in the test field:

$$\frac{\Delta I}{I} = \frac{f_2}{a} \int_{\zeta_1}^{\zeta_2} \frac{1}{n} \frac{\partial n}{\partial y}\, dz. \tag{3.24}$$

For gases one usually has $n \cong 1$, and with the Gladstone–Dale relation [Eq. (3.6)], Eq. (3.24) reduces to

$$\frac{\Delta I}{I} = \frac{K f_2}{a} \int_{\zeta_1}^{\zeta_2} \frac{\partial \rho}{\partial y}\, dz. \tag{3.24a}$$

With the present arrangement we measure the y component of the refractive index gradient. Turning the knife edge (and the linearly extended light

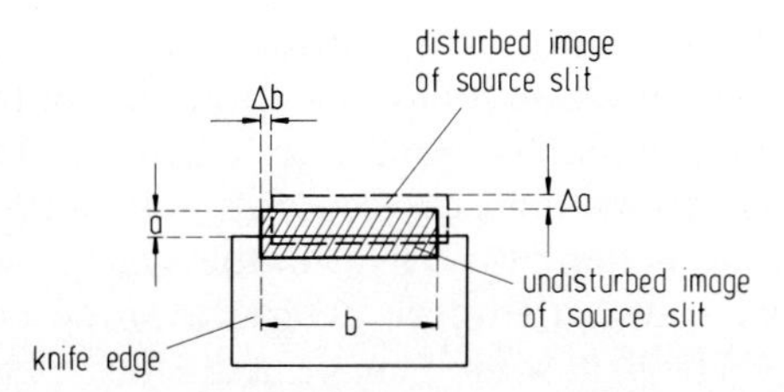

Fig. 3.12 Shift of slit-shaped light source image in the plane of the schlieren knife edge (focal plane of the schlieren head).

source) by 90° around the optical axis makes the system sensitive to the displacement Δb (Fig. 3.12), and one measures the respective x component of the gradient.

From Eq. (3.24) it follows that, for a given refractive index field, the relative change in light intensity and therefore the contrast on the photographic film will be larger the smaller the ratio (a/f_2). If one supposes, that a relative intensity change of $\Delta I/I = 10\%$ is still detectable, the smallest deflection angle which could be measured in a schlieren system is $\varepsilon_{\min} = 0.1\,(a/f_2)$. There is usually no possibility in a given arrangement to change during experiments the focal length f_2 of the schlieren head, and f_2 will be fixed for all test conditions. However, there are also several, mainly three, arguments, that the knife edge aperture a cannot be reduced arbitrarily, and it is for these reasons that the sensitivity range of the schlieren system is limited. Since the illumination of the photographic plate decreases with a, one factor determining a minimum value of a is the speed of the photographic material. While this is no severe limitation, another one results from the demand that the system should be capable of measuring light deflections in both directions (i.e., positive and negative values of ε). In the schlieren system of Fig. 3.11 positive ε generates an increase of the light intensity, negative ε an intensity loss. The greatest possible intensity change for negative ε is absolute extinction; if $\varepsilon_{\max}$ designates the greatest deflection angle in a given density field, then

$$\Delta I = -I \qquad \text{for} \qquad \varepsilon = -\varepsilon_{\max}\,.$$

With Eq. (3.23) one obtains the following condition for the smallest possible a:

$$a_{\min}/f_2 \geqq \varepsilon_{\max}\,. \tag{3.25}$$

3.2.2.2. *Limitations in Resolution of the Teopler System and Quantitative Evaluation*

The most severe restriction on reducing the aperture a results from the influence of diffraction. This effect is not included in the theory of geometric optics, which had been used earlier to describe the schlieren principle. The influence of diffraction on the formation of the schlieren image can be studied independently [see, e.g., Schardin (1942), Suchorukich (1968), or Hosch and Walters (1977)]. Diffraction effects are the more prominent the more coherent the light in the system, and in most cases, the use of a laser source therefore deteriorates the characteristics of a schlieren system. The light source image in the focal plane of the schlieren head is a diffraction pattern with a central maximum surrounded by a system of minima and maxima of higher orders. When white light is used, the light source image is to a certain degree out of focus, because of the overlapping and blurring of the diffraction orders for the different wavelengths. The knife edge is cutting off part of the diffraction pattern, this part being different for the different deflection angles, which the light experiences in the object field. As a consequence, the imaging of the object onto the observation plane is disturbed, and at the same time, the observed intensity change is not accurately described by Eq. (3.24).

A number of authors have investigated the disturbing effects of diffraction for particular test objects of given shape and density distribution. For example, Schardin (1942) studied the diffractive disturbance of the schlieren image of a circular regime with constant deflection ε inside the circle and with a constant density field outside. The object of the study was to analyze, for this particular case, the smallest possible value of a so that the angle ε is determined correctly from the measured intensity change $\Delta I/I$; or, to determine the smallest deflection angle that can be measured by applying Eq. (3.24). Schardin's result is

$$a_{\min}/f_2 \geqq \lambda/2d, \tag{3.26}$$

where d is the diameter of the circular regime. This is a result typical for the application of diffraction theory, which states that the diffractive disturbance is the smaller (or here: the resolution is the higher), the smaller the wavelength λ and the larger the field of interest. The minimum of the deflection angle which can be measured, $\varepsilon_{\min}$, follows from Eq. (3.26) as

$$\varepsilon_{\min} = \frac{a_{\min}}{f_2}\left(\frac{\Delta I}{I}\right)_{\min},$$

and by setting $(\Delta I/I)_{\min} \cong 0.1$. It should be made clear that these results apply only to the specific configuration considered by Schardin. It has been tolerated here that the circular disturbance is not properly imaged in the recording plane. The image of the circular regime, in this extreme case when $a = a_{\min}$, is out of focus to an extent twice the size of the undisturbed image. In other words, the schlieren system is operated at its limits of sensitivity so that the density gradient is measured correctly, but the spatial resolution of the test field is lost completely. The opposite case, high spatial resolution but moderate sensitivity with respect to the measured density distribution, has been studied by Hosch and Walters (1977).

The effects of diffraction can be included directly in the schlieren analysis, if the light modulation in a schlieren system is described by means of Fourier transforms instead of being represented by geometric optics relations. Such an analysis has been performed by Véret (1970) for a light wave that experiences a phase shift $\varphi(x, y)$ in the test field. The relative intensity change or contrast of the schlieren picture is given by

$$\Delta I/I = \varphi^2/(\eta^2 + \beta), \tag{3.27}$$

where the transmission factor η describes the attenuating effect of the knife edge, and the additive term β accounts for the effects of diffraction. We see again, that in the diffraction-free case ($\beta = 0$) the contrast or sensitivity becomes infinite if the transmission is reduced to zero ($\eta = 0$ or $a = 0$), and that the presence of diffraction ($\beta \neq 0$) causes the contrast to remain finite even for $\eta = 0$. Véret (1970) proposes a value of $\beta_{\min} = 5\%$ for a well aligned system. This allows one to determine a smallest detectable phase change $\varphi(w, y)$ as a measure of the sensitivity. This result is somewhat contradictory, since we know from Eq. (3.24) that the schlieren method responds to changes of the density gradient and not to the changes of optical phase or absolute density. These two results must be interpreted in such a way that Eq. (3.24) describes the form of the disturbance necessary for being resolved with the schlieren system, whereas from Eq. (3.27) one derives the smallest amplitude of the disturbance, which can just be detected.

The estimates derived here for the sensitivity limits are theoretical results which apply to an ideal optical system. Experimental values obtained [e.g., by North and Stuart (1963), Philbert (1964), and Véret (1970)] are relatively close to these theoretical limits. The real systems suffer from a number of optical errors or aberrations, which, however, can either be eliminated or at least minimized. These aberrations are introduced by the used lenses or spherical (or parabolic) mirrors, particularly if the mirrors have to be used in an off-axis configuration (see Section 3.2.2.3 and Fig. 3.14). At small off-axis angles coma is the most significant aberration. The two-mirror *Z* configuration (Fig. 3.14) cancels the coma in the knife-edge plane. Astigmatism can either be corrected by a system of plano-cylindrical lenses (Prescott and Gayhart, 1951), or by using a vertical knife edge in the tangential focal plane or a horizontal knife edge in the sagittal focal plane (Hosch and Walters, 1977). The latter correction limits the choice for the direction of the density gradient to be measured.

If all the aberrations and sources of optical errors are minimized, the recorded schlieren image (e.g., Fig. 3.13) can, in principle, be evaluated for determining the distribution of the deflection angle $\varepsilon(x, y)$ or, after a respective integration, the distribution of the fluid density $\rho(x, y)$. (As usual, the x, y plane is the recording plane normal to the optical axis.) A number of attempts have been made in the past for such a quantitative evaluation. When a schlieren photograph is taken, the recorded intensity pattern $I(x, y)$ is influenced by the characteristics of the photographic film

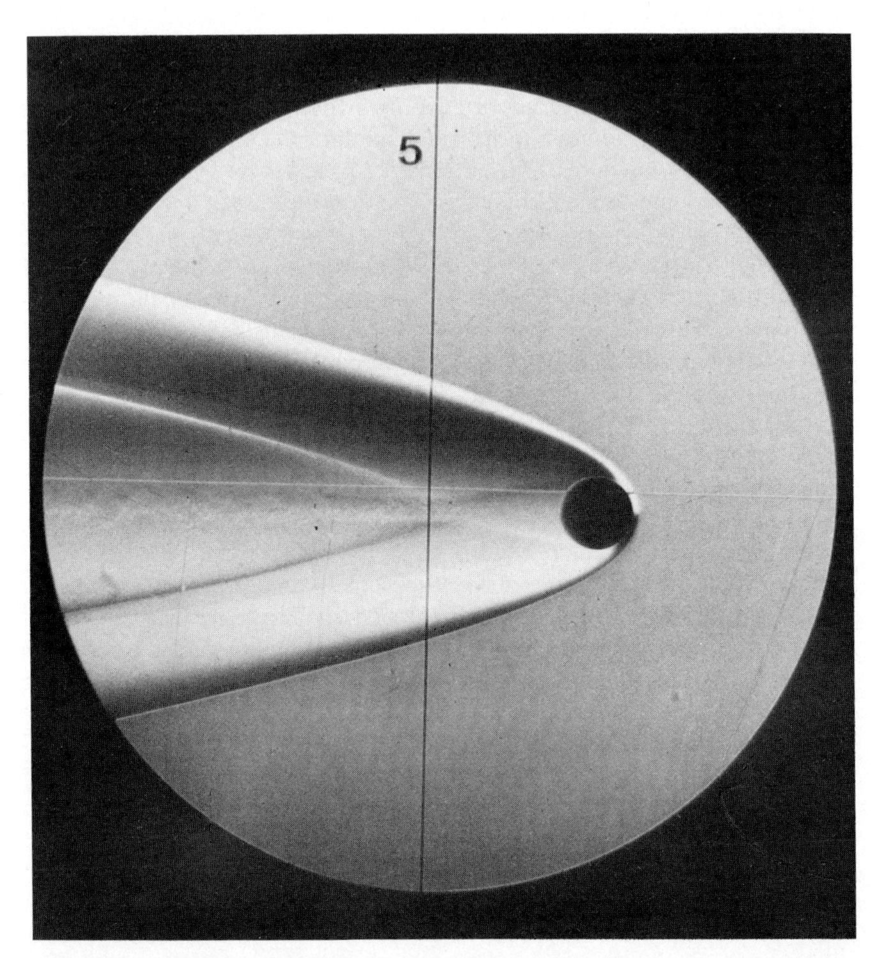

Fig. 3.13 Schlieren photograph of the flow field around a sphere flying at hypersonic speed. This record was taken with a horizontal knife edge, and, since the flow field is axisymmetric, the same density change (compression or expansion, respectively) appears with opposite change in light intensity on either side of the flight axis. (From A. Stilp, Ernst–Mach–Institut, Freiburg, Germany.)

material and the developing process. Suppose that the film is used in its linear sensitivity range and that a linear developing process is applied. Then, the difference in the photographic density, ΔD, at two points (x_1, y_1), (x_2, y_2) on one schlieren photograph is given by

$$\Delta D = \text{const}\ \log[1 + (\Delta I/I)]. \tag{3.28}$$

By measuring ΔD with a densitometer one can determine $\Delta I/I$ and thereby calculate ε, η, or ρ from Eqs. (3.23) and (3.24). The constant in Eq. (3.28) depends on the photographic process and can be determined if one knows ε or $\Delta I/I$ in one point of the field (which is usually not the case). An alternative approach for quantitative evaluation is to include in the object field a number of "standard schlieren", which may take the form a series of glass wedges of known deflection angle. The images of the "standard schlieren" on the film are then used as reference values.

The densitometric evaluation of the schlieren photography can be automated, and by scanning the photograph it is possible to identify curves of equal photographic density, which are available for the further numerical evaluation (Corcoran, 1967). Such curves can easily be generated if the record is taken by a television camera with subsequent digitization of the image (Stanic, 1978). Decker *et al.* (1985) report on a computer simulation of schlieren pictures by optical ray tracing; the comparison of the so calculated and the visualized structures is another means for analyzing the schlieren record. Despite of all these attempts for quantitative evaluation, the schlieren method today is essentially used for qualitative investigations of density fields, but with a much higher degree of resolution than the shadowgraph. There are mainly two reasons for the restriction of the schlieren method to qualitative purposes. The measurement of (photographic) contrast or shades of gray is never very precise, and a schlieren setup can easily be converted into a system being more appropriate for quantitative density measurements (see, e.g., Sections 3.2.3, 3.2.4, and 3.3.3).

3.2.2.3. *Modifications of the Toepler System*

A wide variety of modifications of the Toepler system has been described in the literature, of which a few will be discussed here. Fig. 3.14 shows the Z configuration in which the lenses depicted in Fig. 3.11 are replaced by two spherical or parabolic mirrors. A system with mirrors is required if the field of view is large, because mirrors can be fabricated with a much greater diameter than lenses. The Z configuration has already been mentioned in connection with the optical aberrations coma and astigmatism.

In order to increase the sensitivity, another arrangement, called the "double-pass schlieren system," uses a conical light beam passing twice

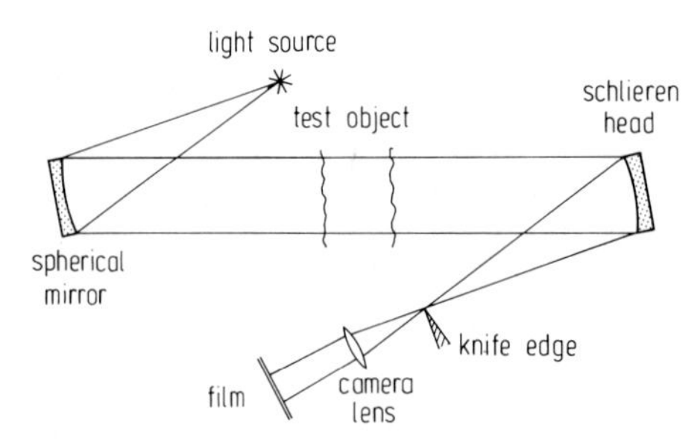

Fig. 3.14 The *Z*-configuration of schlieren system with two spherical (or parabolic) mirrors.

through the test field (Fig. 3.15). Such a system is often applied in ballistic ranges to study the wake of hypervelocity projectiles under extreme low density conditions (Slattery *et al.*, 1968; Royer and Smigielski, 1968). The divergent–convergent light beam, including the spherical mirror, can be totally kept in an evacuated tube to protect the system from the thermal turbulence of the laboratory room. If a prism is used for separating the incident and reflecting beams (Fig. 3.15a), the optical path is again "off-axis;" objects in the test field appear with a double image; and an incident ray traverses the field at a slightly different position from the reflected

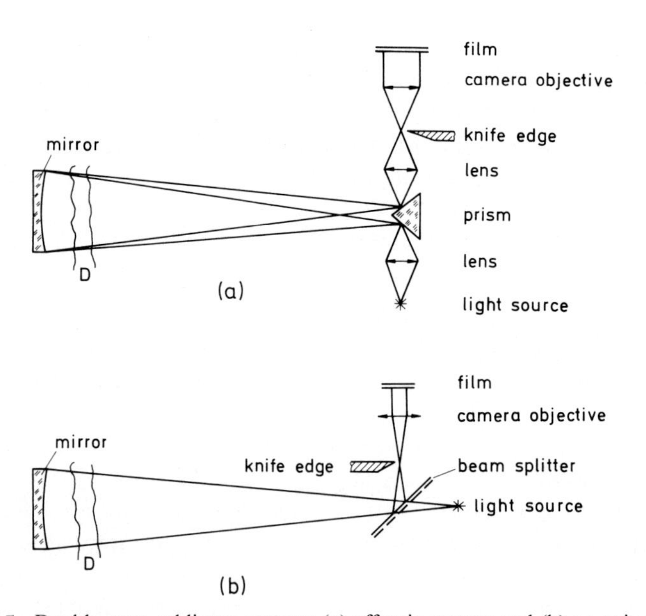

Fig. 3.15 Double-pass schlieren system: (a) off-axis system and (b) on-axis system.

ray. This disadvantage can be overcome by using a beam splitter (semireflecting mirror), and it is only in this case, that the result of the double-pass system is to really double the effective width of the test field and so to double the sensitivity (Fig. 3.15b); at the same time such an arrangement is free of coma and astigmatism. Houtman and Meyer (1984) describe a schlieren-like system, which more than doubles the schlieren sensitivity. By means of Fabry–Pérot mirrors a parallel laser beam is made to pass several times, back and forth, through the density field.

Further modifications of the Toepler system concern the shape and function of the knife edge. If the aperture a of the conventional knife edge is reduced in order to increase the sensitivity, Eq. (3.25) might be violated. Then regions in the schlieren image, where the deflection angle ε is negative and where the absolute value exceeds a certain value ε, are obscured completely, whereas, the regions with positive ε exhibit a distinguishable contrast, also beyond this limit ε. Another observation is that the eye resolves changes in contrast or intensity much easier for an elevated intensity level (positive values of ε), than in the case of a low average illumination (negative ε). Modifications of the knife-edge geometry, the "double knife edge" and the "circular cut off" method, improve the schlieren sensitivity in the increasing intensity range (Stolzenburg, 1965). These modifications are also known as "dark field" methods, because the conventional "single" knife edge is replaced by an opaque diaphragm or stop cutting off symmetrically all but a very small fraction of the undisturbed light source image in the focal plane of the schlieren head. The undisturbed field, therefore appears at a very low intensity (dark field). The double knife edge is applied when a slit-shaped light source is used (Fig. 3.16); positive and negative values of ε_y (if the knife edge is horizontal) are visualized with the same contrast, but can no longer be distinguished. In the circular cut-off system all deflection angles of the same absolute value are producing the same contrast, regardless of their orientation in the x, y plane. An opaque schlieren stop can be made by taking photographs of the light source image at different exposure times and then using the appropriate negative (Stanic, 1978). Vasil'ev and Otmennikov (1978) have performed a theoretical investigation of the schlieren sensitivity for such dark field systems.

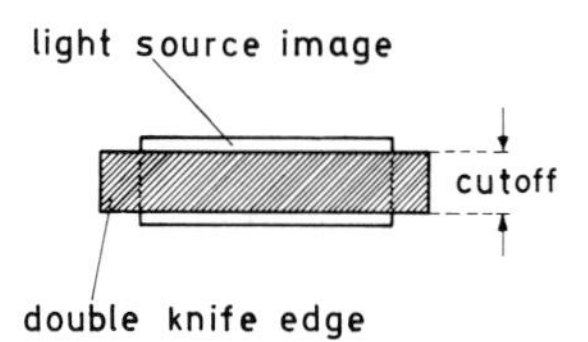

Fig. 3.16 Double knife edge.

The replacement of the conventional knife edge by a filter having a gradual variation of light transmission may reduce the unwanted diffractive effects. Such a graded filter can be fabricated by evaporating a metal film onto one half of a glass plate; within a penumbra zone the coating thickness varies smoothly, so that the transmission of the plate (or its reflectivity) changes gradually from zero to one (Kent, 1969). A variety of other elements replacing the conventional knife edge have been investigated, for example, a quartz prism that rotates the plane of polarization by different amounts depending on the angle of light deflection (Andrews and Netzer, 1976), or a lens with uncorrected spherical aberration and short focal length (Dixon, 1982).

For some of the modifications it is hard to say whether we should still classify them as a schlieren system. We shall restrict the discussion to such devices, which affect the intensity of the light in the focal plane of the schlieren head and not its phase. The latter case is treated in Section 3.4. A system of the former type has been described by Knöös (1968) and is actually based on an earlier method, sometimes referred to as the Thovert–Philpot–Svensson "crossed-slit method" (for details see, e.g., the article by Schardin, 1942). Knöös' system makes use of the astigmatic aberration in an off-axis arrangement containing one spherical mirror (Fig. 3.17). The astigmatic vertical and horizontal linear images of the point light source are separated by a small distance along the optical axis. An inclined, thin wire is placed in the plane of the first (vertical) light-

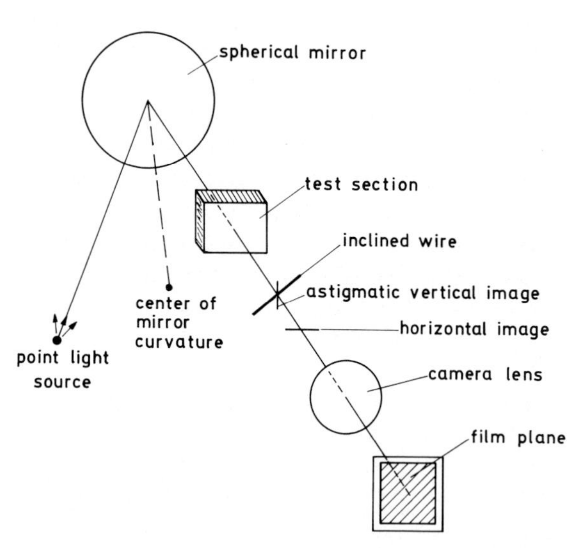

Fig. 3.17 Cross-slit method according to Knöös (1968).

source image, and the shadow of this wire is focused onto the film plane, where it appears as a black fringe. The refractive deflections in the test field produce a fringe shift Δd, which for the case of a one-dimensional distribution of the density gradient can easily be related to the deflection angle by $\Delta d = k\varepsilon(\mathrm{x})$; $k =$ const. for a given optical system. The sensitivity of this system depends on the size of the light source; the wire diameter must be greater than the thickness of the focal image line; whereas, the smallest observable fringe shift is of the order of the wire diameter. Knöös refers to a smallest observable value of $\varepsilon = 10^{-4}$ rad. A reduction of the wire size results in a pattern of diffraction fringes, thereby limiting the resolution. Grossin *et al.* (1971) report on a device where a grid is placed between the two astigmatic images of the light source. In general, the result of this method is to produce, as the measuring signal in the field of view, a fringe whose deformation is a measure of the deflection angle. For the purpose of quantitative evaluation such a signal form is more handy than the photographic contrast of a conventional schlieren picture.

The Toepler method produces a two-dimensional image of a three-dimensional flow field, and the computation of the density distribution $\rho(x, y, z)$ from the recorded intensity pattern is a more or less unsolvable problem. Apart from other visualization methods, which offer a more convenient way for three-dimensional resolution, attempts have been made to modify the schlieren system for the investigation of three-dimensional objects. The "sharp-focusing schlieren system", described by Kantrowitz and Trimpi (1950), uses an extended light source or a plane distribution of multiple point sources and multiple knife edges, one for each source. A plane S in the test section, perpendicular to the optical axis, is focused onto the film plate by means of the camera lens. Through any point of the plane S passes one ray of each source, and this set of rays converges again in any corresponding image point on the film. Along their paths each of the rays passes through different points of the test field and thereby undergoes a deflection different from that of the other rays. Only the deflection from the common point in S is common to all rays of one set; but all the rays (i.e., one from each source) contribute to the intensity of illumination of the common image point on the film or screen. It is believed that the intensity components, which result from the common deflections in the plane S add in the image; whereas, the other components resulting from the deflections along the different path segments will average out; the intensity pattern of this schlieren image reflects, therefore, on a basic background level, the density distribution at the plane S of common points.

The application of this system requires a rather complicated procedure. In order to obtain a three-dimensional density distribution it is necessary to change the plane of focus S several times during one experiment and eventually to realign the multiple knife edges. Buzzard (1968) has shown that some of these difficulties can be overcome by the use of holography.

Any optical visualization system can be combined with a holographic set-up, for example, for the purpose of gaining practically unlimited time for performing the optical experiment. If one takes a hologram through the test field one may freeze the instantaneous information of the refractive index field. Figure 3.18 shows the principal arrangement for taking such a hologram of the phase object and for the subsequent schlieren observation of the reconstructed object field (Trolinger, 1974). If the "sharp-focussing schlieren system" is applied, the procedure necessary for this system can now be performed carefully to obtain a maximum of resolution. Rotem *et al.* (1969) report on a modification of this system using colors.

Techniques analogous to those optical probes, which have been mentioned at the end of the section on shadowgraphy also exist for the case of the schlieren methods: a single, unexpanded laser beam is transmitted through an unsteady density field and received by a photomultiplier,

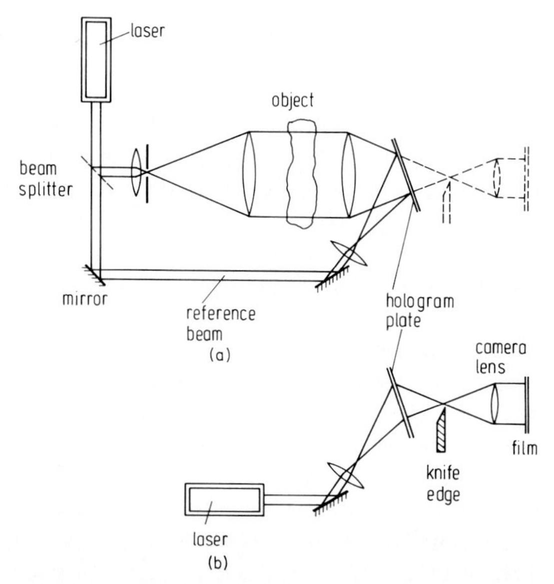

Fig. 3.18 Arrangement for taking a hologram of a transparent object field with refractive index or density variations (a) and postponed schlieren visualization of the reconstructed object field (b).

which records the signal continuously in time. Here, however, the beam is passed over a knife edge before entering the photomultiplier. Such devices can be used for studying the characteristics of turbulent density fields (Koziak, 1970; Davis, 1972) or unsteady phenomena in shock tubes (Kiefer and Lutz, 1965; Bander and Sanzone, 1974; Kiefer *et al.*, 1981). Spatial resolution has additionally been introduced into these narrow-beam techniques by the "crossed-beam correlation method" of Fisher and Krause (1967). This technique, in principle similar to the "sharp-focusing schlieren system," involves the use of two narrow light beams, which traverse the turbulent flow field in two perpendicular directions, and which can subsequently be combined to form one resultant beam arriving at a light detector. This detector analyzes the intensity fluctuations of the resultant beam. Each beam alone reflects the integral of flow properties along its path. If the covariance (time-averaged product) of the two detected signals is formed, only the "correlated" fluctuations near the intersection regime of the two beams will yield a finite-averaged product; whereas, the (uncorrelated) fluctuations, which occur at a sufficient distance from this intersection point, will average to a value of zero.

Ultimately, it should be mentioned that, in another type of optical probe, the schlieren effect is used for determining the velocity of a density disturbance by measuring the Doppler shift of the deflected (laser) light (Schwar and Weinberg, 1969; Merzkirch and Erdmann, 1974).

3.2.2.4. Color Schlieren Systems

Since the eye is more sensitive to changes in color than to changes in shades of gray, this new dimension has been introduced into the schlieren technique in various forms. One approach is to replace the knife edge by a filter, which consists of several parallel, transparent, colored strips made, e.g., of cellulose acetate sheets (Kessler and Hill, 1966) or of commercially available gelantine filters (Smith and Waddell, 1970). Most often three colored sheets are used, and this "tricolor filter" is arranged to be parallel to the light-source slit, the width of the central filter section being approximately equal to that of the slit image. The choice of the colors for the three strips depends on the appearance and the discrimination by the eye, as well as on the color sensitivity of the film material; the three color sections should have approximately the same transparency, and a combination of red, blue, and yellow seems to yield the best contrast.

The tricolor filter is just one approach among a great variety of possible color filters, which have been reviewed by Settles (1982). A few of them, which are basically different are mentioned here. The method of Holder and North (1952) employs a dispersion prism in front of a source of white light located at the focus of the first spherical mirror (Fig. 3.14). A color spectrum is thus displayed in the focal plane of the schlieren head. The knife edge is replaced by a narrow slit, that is adjusted to pass only one color, usually yellow in the undisturbed case. With a variable density gradient in the test field, light of a different color traverses the slit. The sequence of the displayed colors is that of the natural spectrum, and one has no choice of selecting a highly contrasted combination as with the tricolor filter. The separation of white light into its spectral color components is achieved in systems described by Maddox and Binder (1971) and by Chashechkin (1985) by means of a diffraction grating. Attempts are made to use this color schlieren device for quantitative evaluation.

While the circular cutoff system is equally sensitive to deflections in all directions, a colored modification of this method (Settles, 1970) discriminates also the direction of the deflection angle in the x, y plane. A white light source is focused on a colored source filter, which consists of four differently colored slits arranged in a square (Fig. 3.19). With the source filter being placed in the focus of the first spherical mirror the light of the four colors overlaps and mixes; in the absence of a flow disturbance the schlieren image is evenly illuminated by light of a single hue. In the place of a knife edge one uses a squared aperture of side length somewhat smaller than the length of the outer edge of the source filter. The direction of a density gradient in the test section determines which color combination can pass through the aperture to illuminate a local region of the schlieren image. The color combination of the source filter is chosen, so that light from adjacent bands of the filter can mix in the schlieren image. The recorded pure colors and color combinations are a measure for the local direction of density gradient in the test section. Settles's system has been used by Stastny and Pekarek (1985) for visualizing the flow through a turbine profile cascade. Howes (1984) has devised another multidirectional filter in which the sequence of colors in radial direction is that of the natural spectrum ("rainbow filter"). Diffraction effects are minimized by this continuous change of colors.

An obvious requirement for any color system is the use of a white or nearly white light source. O'Hare and Trolinger (1969) have shown how

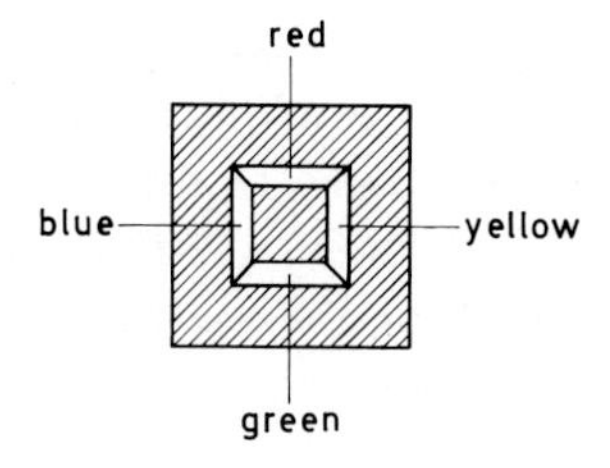

Fig. 3.19 Colored source filter for directionally sensitive schlieren system. (From Settles, 1970.)

color schlieren can be produced if the schlieren apparatus is applied to the reconstructed object field in a holographic setup where the illumination is monochromatic (Fig. 3.18). This procedure, however, is very sensitive to disturbances by diffraction.

3.2.2.5. *Applications of the Schlieren Method*

A very great variety of applications exists for any field listed in Section 3.1. The number of applications for high-speed flows, gas dynamics, and combustion is immense, and as we do it also in other cases, a few representative applications in areas other than high-speed flows will be mentioned here. Convective heat transfer has been observed in air and in various liquids as well (Carey and Mollendorf, 1978). Jet mixing (Ishikawa, 1983) and the associated formation of coherent structures (Davis, 1982) and the generation of jet noise (Heavens, 1980) are sensitive objects for schlieren visualization. Hsia *et al.* (1984) have shown that by properly selecting the reservoir temperatures in such jet studies one may enhance the conditions for visualizing the flow structures.

The phase distribution of internal waves in density stratified fluids, usually saltwater systems, has been visualized with conventional (Mowbray, 1967; Stevenson, 1973) and with color schlieren photography (McEwan, 1983). Gärtner (1983) has performed amplitude measurements in an internal wave by visualizing in the brine the deformation of vertical traces of sugar due to the wave motion (Fig. 3.20).

3.2.3. *Deflection Fringe Mapping*

As we had discussed in connection with the method of Knöös (1968), the distortion of an originally straight fringe is a signal more appropriate for quantitative evaluation than the measurement of photographic contrast. Many attempts have been made to modify the shadow as well as the schlieren method to obtain flow pictures with a superimposed network of fringes, whose distortion is a measure of the light deflection in the test field. Weinberg (1963) named these techniques ''deflection mapping,'' and some of them are also referred to as ''Ronchi methods.'' By superimposing two systems of fine fringes one produces a ''moiré'' pattern, another name under which some of the methods to be discussed here are known.

The color schlieren system using a tricolor filter is a simple form of such a deflection mapping system. Each line separating two colors in the recorded picture is a curve of constant deflection angle ε_y, if the edges of the filter are parallel to the x direction. Instead of the filter one may use a grid of alternate transparent and opaque strips of equal width (Kogelschatz and Schneider, 1972). The spacing of the strips is chosen so that

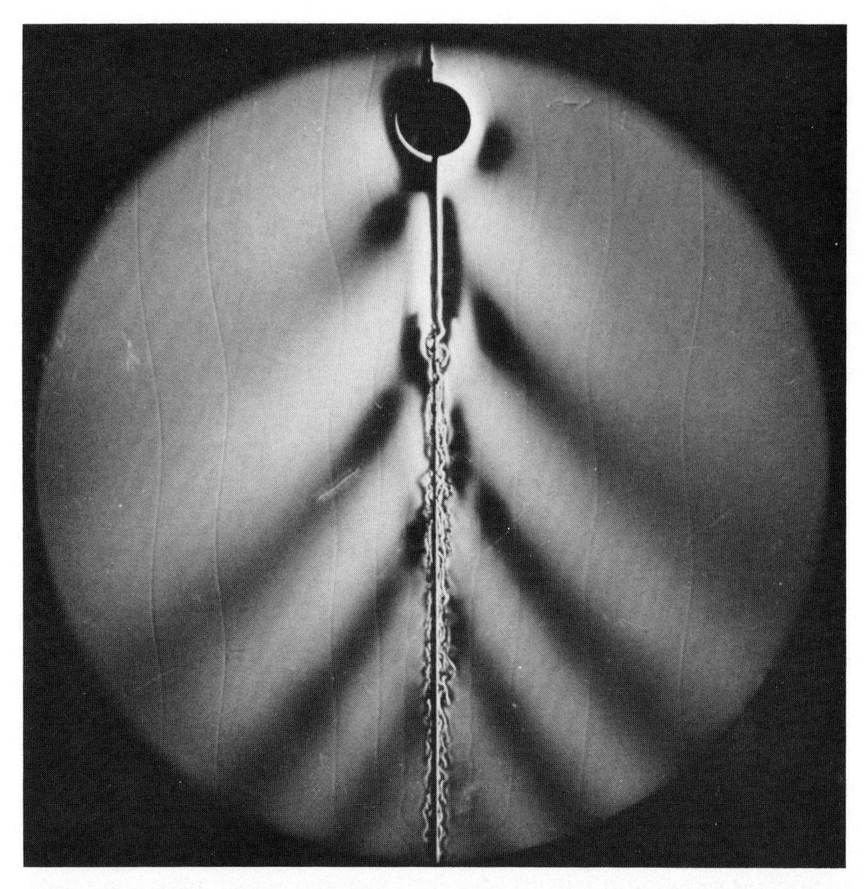

Fig. 3.20 Toepler schlieren visualization of the internal gravity wave behind a cylinder moving in upward direction through a density stratified saltwater system. Also visible are the traces of minute sugar crystals, which had been dropped into the brine immediately before the cylinder was started from its initial position. These traces are deformed according to the local amplitude of the wave. (From Gärtner, 1983.)

the width of the schlieren light source image is equal to or less than the width of the central aperture, and with a uniform test field the recording plane will appear evenly illuminated. With small deflection angles the system works like a double-knife edge arrangement. With stronger density gradients in the flow field (Kogelschatz and Schneider applied the system to a high current arc!), the rays might be deflected in the plane of the grid by more than one grid period, so that dark and bright fringes appear in the recording plane, each of which being a curve of constant (range of) deflection angle. The precision in determining the deflection angle depends on the focal length of the schlieren head and on the fineness of the strip system. Diffractive disturbances, however, increase with the fineness of the grid.

The described system is characterized by a field of view free of fringes for a uniform test object. When the grid is shifted from the focal point of the schlieren head along the optical axis toward the test section (Fig. 3.21), an equidistant fringe pattern appears in the recording plane, even in the undisturbed case. If a density disturbance occurs in the flow, the fringes are no longer straight and parallel. The observed fringe distortion (fringe shift) is proportional to the amount of light deflection. It is apparent that the two systems with the grid either in the focal plane or out of the focal plane of the schlieren head are somehow equivalent to the "infinite fringe width" and "finite fringe width" alignments of optical interferometers, particularly of shearing interferometers (see Section 3.3.3). If the fringe shift is measured normal to the undisturbed fringe direction, here assumed parallel to the x axis, one may relate the absolute shift Δs_y to the deflection angle ε_y, according to Holder and North (1963), by

$$\Delta s_y = \varepsilon_y \left\{ f_2 \frac{f_c}{g} + (1 - f_2) \frac{g}{f_2} \right\}. \tag{3.29}$$

The notation in this equation follows from Fig. 3.21. The fringe shift does not depend on the spacing of the strips in the grid. The case $g = 0$ is equivalent to a schlieren system with infinite width of the fringe.

The grid can also be placed on the left side of the schlieren head, or on the opposite side of the test object (i.e., into the incident light beam before it enters the flow field). The grid can be different in shape, showing, for example, a regular pattern of concentric circles or squares. In principle,

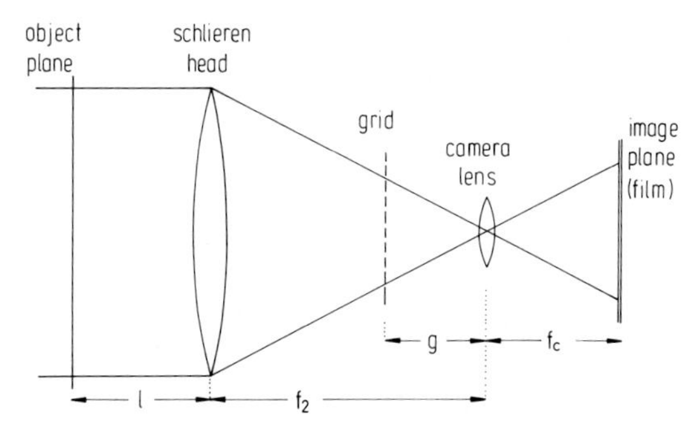

Fig. 3.21 Deflection mapping system with a grid producing fringes whose distortion is measured in the image plane.

one observes through the test field a pattern of known geometry, which is now distorted due to the refractive deflection of the light rays in the density field. Equations like Eq. (3.29) can be derived by geometric ray tracing. A great number of such arrangements have been used, for both qualitative and quantitative purposes, and mostly applied to flow situations, which are governed by relatively strong diffraction, for example, flames (Fig. 3.22), shock waves (Thompson *et al.*, 1985), or stratified saltwater systems (Mowbray, 1967; Thorpe, 1973).

If one superimposes two fine, straight grids of equal frequency, but rotated at a certain angle θ, a system of fringes becomes visible, which usually is referred to as a moiré pattern. Kafri and his co-workers have used such moiré methods for measuring the refractive deflection in a number of different flow situations (Kafri, 1981; Keren *et al.*, 1981; Stricker and Kafri, 1982; Bar-Ziv *et al.*, 1983; Kafri *et al.*, 1984). Two parallel gratings ("Ronchi rulings") with a distance ("pitch") $\mathbf{s}$ between the grooves are inserted into the light path behind the test field. The gratings have an axial distance Δ, which must be chosen so that a good clarity of the moiré fringes is achieved. In the reported experiments values $\mathbf{s} = 0.18$ mm and $\Delta = 290$ mm had been used, and the moiré pattern was produced on a screen immediately behind the second grating. The distance S between the undisturbed, parallel moiré fringes (see Fig. 3.23) is

$$S = \frac{\mathbf{s}}{2\sin(\theta/2)}. \tag{3.30}$$

The refractive deflection of light normal to the direction of the undisturbed moiré fringes is visualized by means of a fringe shift ΔS. If the undisturbed moiré fringes are assumed to be parallel to the x axis, then

$$\Delta S = \varepsilon_y(\Delta/\theta) \tag{3.31}$$

with the usual notation that ε_y is the component of the light deflection angle in y direction. The quality of the fringe pattern, which can be achieved is very similar to that of an optical interferogram (Fig. 3.24).

3.2.4. Deflection Mapping by Speckle Photography

It has been explained in Section 2.2.4 that laser speckle photography is a method for measuring in-plane displacements and deformations. It was shown first by Köpf (1972) and by Debrus *et al.* (1972) that the speckle method can be modified so that the light deflection in refractive index fields can be determined quantitatively. This method is therefore compa-

Fig. 3.22 Deflection mapping system using a grid of equidistant straight wires. The grid pattern is distorted due to the plume raising from a candle flame. (From W. Erdmann and W. Merzkirch, Ruhr–Universität Bochum.)

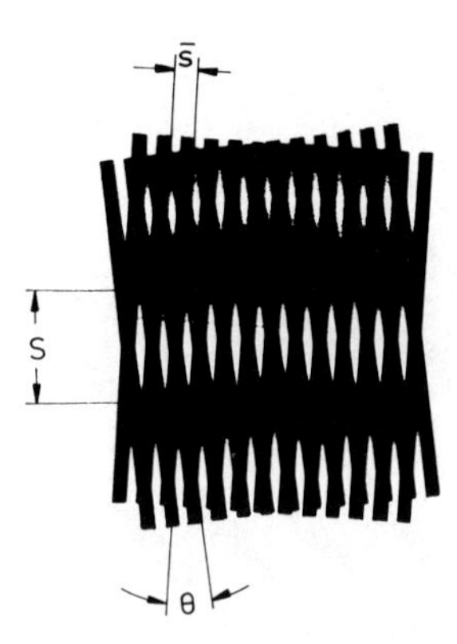

Fig. 3.23 Generation of a moiré pattern by the superposition of two grids rotated at an angle θ. Definition of "pitch" s and fringe width S.

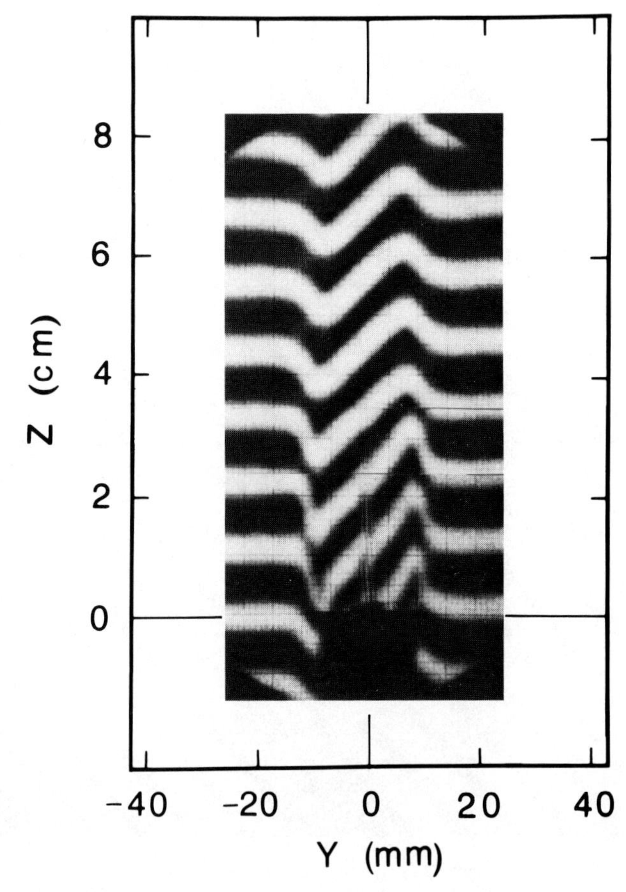

Fig. 3.24 Moiré pattern of a CH_4–air flame. (From Bar-Ziv *et al.*, 1983. Copyright, Optical Society of America.)

rable to the quantitative schlieren and deflection mapping systems or to shearing interferometers (Section 3.3.3), but we shall see that these laser speckle techniques have many advantages over the former methods, and possibly could replace most of them.

The optical arrangement used by Köpf (1972) is shown in Fig. 3.25. An expanded laser beam illuminates a plate of ground glass. This ground glass is imaged onto a photographic plate, and a first photographic exposure is taken in the absence of the test field. A "speckle pattern" is thereby recorded; it is caused by the random distribution of nonuniformities in the ground glass. After this first exposure, the flow field with varying density or refractive index distribution is introduced in the optical system as seen in Fig. 3.25. A second exposure is taken on the same photographic plate. Due to the light deflection in the test field, the speckle pattern recorded in the second exposure is different from that in the first exposure. Individual speckles appear displaced by a distance Δ, if the refractive ray in the imaging system is deflected by an angle ε, with $\varepsilon \cong \Delta/l$, where l is the distance of the test field from the ground glass. The photographic double exposure with the two recorded speckle patterns therefore contains a, more or less, continuously distributed information on the light deflection angle ε in the field of view (x, y).

We could regard this double-exposure technique as a deflection mapping method with the ground glass being a grid of very fine structure, although the formation of light is really a process of multiple interference of light scattered by the ground glass. The deformation of the grid can hardly be recognized with the eye, and for measuring or displaying the distribution of light deflection one applies to the developed double exposure ("specklegram") one of the reconstruction methods explained in

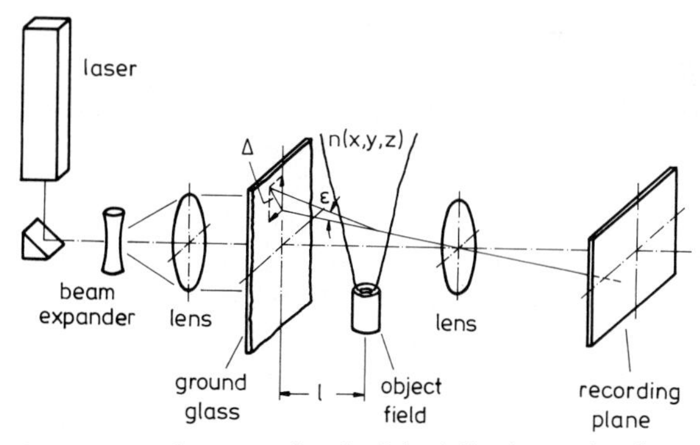

Fig. 3.25 Arrangement for measuring the light deflection angle ε by means of speckle photography. (After Köpf, 1972.)

Section 2.2.4. Since the "point-by-point" reconstruction method (Fig. 2.27) provides the displacement $\Delta(x, y)$ by magnitude and direction, such an evaluation yields from one specklegram the two deflection angles $\varepsilon_x(x, y)$, and $\varepsilon_y(x, y)$ at the same time, in contrast to the Toepler schlieren system, deflection mapping, and shearing interference methods, all of which can measure, in one single experiment, only one component of the deflection angle or the density gradient, respectively. In addition to that, the deflection angle $\varepsilon(x, y)$ can be determined with this laser speckle technique virtually in any desired point (x_i, y_i) in the field of view, that is, a very dense distribution of the measurable signal is available, and the availability of the signal in a point (x_i, y_i) is not dependent, for example, on the existence of an interference fringe in that point.

The point-by-point evaluation can be automated, as already discussed in Section 2.2.4 (see Fig. 2.30), and the data $\Delta(x, y)$ may be stored for further processing in a computer. This is of interest for determining the density or its gradient via Eq. (3.16b), and for applying an Abel inversion or tomographic methods if a three-dimensional density field has to be reconstructed. The automated data reduction is of particular significance in connection with one possibility which the speckle technique offers in contrast to the "conventional" optical methods: Wernekinck *et al.* (1985) have verified that quantitative measurements of $\varepsilon(x, y)$ can be performed even for a turbulent flow with fluctuating density. Such a measurement requires the use of a very short light source (e.g., a ruby laser). One obtains then, for one particular instant of time, the instantaneous distribution of $\varepsilon(x, y)$ in a whole field of view; and since this data is very densely distributed and a large number of data points is available, this data can be statistically analyzed in analogy to the procedure proposed by Uberoi and Kovasznay (1955), see Section 3.2.1. As a first step, one may calculate the correlation function of the deflection angle $\varepsilon(x, y)$; and if isentropic turbulence can be assumed it is possible to determine the correlation function of the three-dimensional, turbulent density field $\rho(x, y, z)$ and other characteristic properties of the turbulent field, which are related to this function. In this procedure, the assumption of isotropic turbulence replaces the missing optical information in the third z direction.

The described speckle technique has been applied for investigating the laminar density fields produced by a flame (Farrell and Hofeldt, 1984), and by natural convection in liquids (Sivasubramanian *et al.*, 1984) and air (Wernekinck and Merzkirch, 1986). If an extended test object causing large values of the light deflection is introduced in the optical set-up of Fig. 3.25, the photographic imaging through this object field is disturbed in comparison to the imaging for the first (reference) exposure. This disturbance results in a decorrelation of the two speckle patterns, and it becomes difficult to reliably analyze the respective system of Young's fringes in the point-by-point reconstruction. For such cases governed by strong light deflection, Wernekinck and Merzkirch (1986) used a modification of the optical arrangement (Fig. 3.26). The test object is placed in front (on the left side) of the ground glass, and it is imaged by means of a lens onto the plane of the ground glass. The photographic plate (recording plane) is focused onto a plane at distance l from the ground glass where corresponding rays from the two exposures (taken with and without test field) appear separated by the displacement Δ. The relation between Δ and ε is again $\varepsilon \cong \Delta/l$. With this arrangement there is practically no upper limit for the extension of the test field in the z direction. The system is equivalent to the measurement of inclination angles.

The sensitivity of the method depends on the size of the individual speckles and the amount of the displacement Δ. In analogy to the discussion in Section 2.2.4, the lower sensitivity limit is reached when Δ is equal to or smaller than the speckle diameter d_p. With the arrangement of Fig. 3.26 there is practically no upper limit for the measurement of large deflection angles.

Besides the quantitative measurement of deflection angles or density, respectively, this speckle photographic technique also allows for a visualization of the density field if the "spatial filtering" reconstruction method is applied to the specklegram (Fig. 2.31). The field of view now shows a system of fringes, which are curves of constant component of the deflection angle. The position of the transparent hole in the Fourier plane (Fig. 2.31) determines which component is visualized in this way. With a small distance of the hole from the center of the Fourier plane one may generate a pattern, which is similar to a schlieren picture (Fig. 3.27), and by using a white light source in the spatial filtering reconstruction, the pattern appears colored.

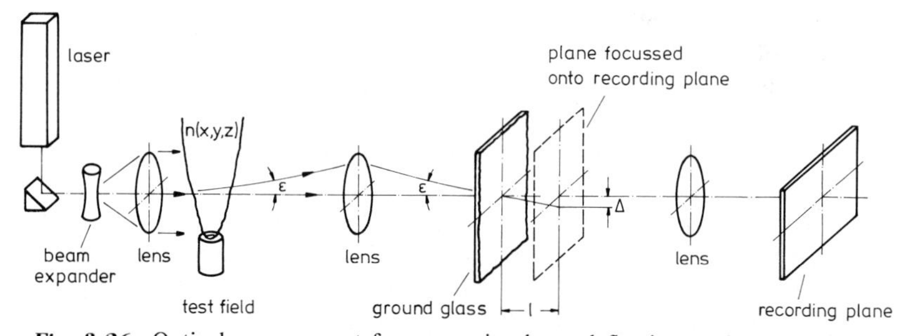

Fig. 3.26 Optical arrangement for measuring large deflection angles ε by means of speckle photography. The distance l must be chosen according to the range of magnitude of ε. (From Wernekinck and Merzkirch, 1986. Published by Hemisphere Publishing Corporation.)

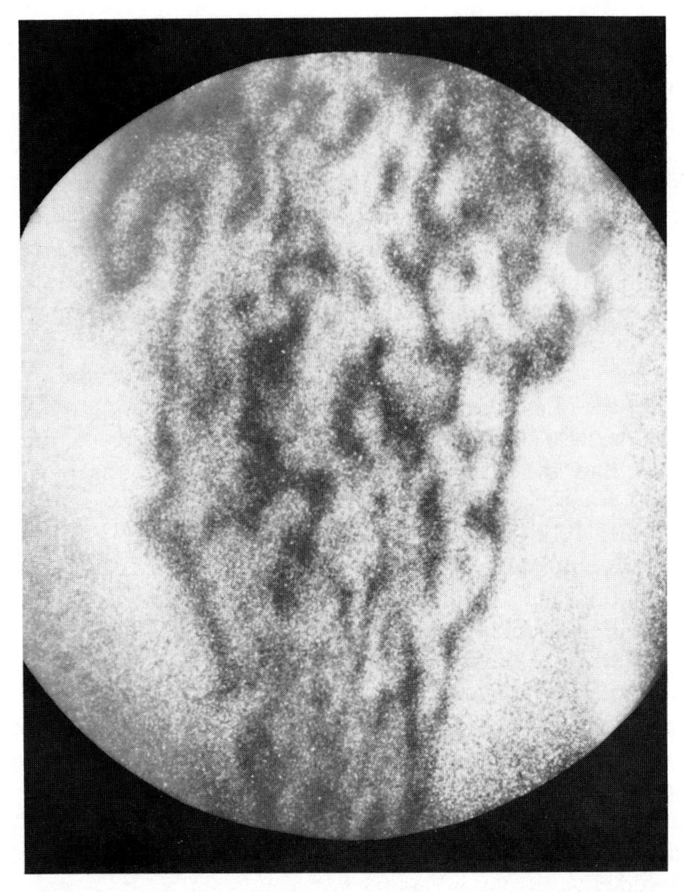

Fig. 3.27 Visualization by means of spatial filtering from a specklegram—turbulent helium jet exhausting into air. (Courtesy of R. Erbeck, Universität Essen.)

3.3. Interferometry

3.3.1. Principles of Two-Beam Interferometry

Along with Fig. 3.3 we had discussed that a flow with variable fluid density represents an optical disturbance, which changes the phase of a transmitted light wave with respect to an undisturbed wave. The development of instruments devised to display such alterations or differences in optical phase has been a great achievement in optics, and has provided instruments known as interferometers. The application of optical interferometers to experimental flow studies began in the late 19th century, when Ernst Mach used a Jamin interferometer to visualize the compressible flow around a projectile flying at supersonic speed. Each interferometric system to be discussed in this section can be regarded as a device in which one visualizes the interference of a wave, which has passed through the test field, with a second wave, which propagates to the recording plane along a different optical path. Hence, we deal with the class of "two-beam interferometers." In terms of optical interferometry, this is already a restriction, because we, more or less, neglect the existence of multiple-beam interferometers. The advantage of this class of interferometric methods is to improve the sharpness of the fringes, but not to increase the sensitivity, and the number of applications of multiple-beam interferometers to fluid mechanical experimentation is very limited (e.g., Frohn, 1967; Harvey, 1970).

The principle of optical interferometry has found application to a great variety of fields besides flow visualization. An extensive literature on interferometry is available, and a great number of instruments has been developed and named in various ways, each of these instruments having a particular advantage for a particular purpose. But we shall see that from these many interferometers only a few are of importance for flow field testing.

We choose the general setup of a two-beam interferometer to have a parallel beam of light passing through the test field. A coordinate system is introduced with the incident light propagating in the z direction (Fig. 3.28). We shall discuss the principle of two-beam interferometry in terms of light rays. Each ray of the incident light beam has a conjugate ray, and the performance of the two-beam interferometers is to measure the phase difference between the two conjugate rays after passage of the test field.

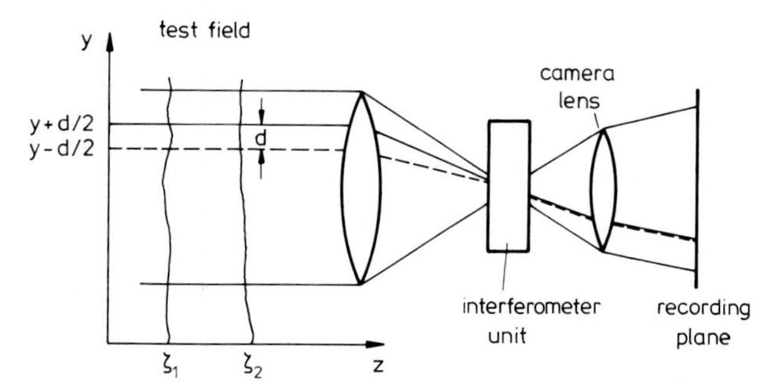

Fig. 3.28 Principal arrangement of two-beam interferometer with parallel light through the test field.

A preliminary assumption is that both rays traverse the test field separated by a distance d. Figure 3.28 is a projection of the y, z plane in which two conjugate rays have the coordinates $y + (d/2)$ and $y - (d/2)$. The test field is bounded by surfaces $\zeta_1(x, y)$ and $\zeta_1(x, y)$. For the following analysis, we assume that these surfaces are planes z = const., that is, ζ_1, ζ_2 are the coordinates where the light rays enter and, respectively, leave the test field. This assumption is no restriction of the analysis, and it can easily be accounted for if the test field does not have plane, parallel boundaries. Behind the test field, a lens or mirror, equivalent to the schlieren head, focuses the light beam. A certain "interferometer unit", which is regarded here as a black box for simplicity, provides that the two conjugate rays either coincide after having passed through this unit, or they are made to intersect in the recording plane. If the conditions of optical coherence are fulfilled, the conjugate rays may interfere with one another and produce an interference pattern in the recording plane.

The difference in optical path length Δl between the two conjugate rays in Fig. 3.28 is

$$\Delta l(x, y) = \int_{\zeta_1}^{\zeta_2} n\left(x, y + \frac{d}{2}, z\right) dz - \int_{\zeta_1}^{\zeta_2} n\left(x, y - \frac{d}{2}, z\right) dz. \tag{3.32}$$

The same values ζ_1, ζ_2 are used in the two integrals because of our assumption that the flow field is bounded by plane walls perpendicular to the z axis. The plane formed by the conjugate rays (here: a plane x = const.) can be changed by rotating the optical setup around the optical axis, which is here parallel to the z axis. Dividing Δl by the wavelength of the used light, λ, yields the optical phase difference $\Delta\varphi/2\pi$, which was used in Eq. (3.17). This equation was derived under the assumption that the light rays follow straight path lines through the flow, or that refractive deflection can be neglected. In Section 3.3.6 we will discuss the case of strong refraction when the latter assumption cannot be made.

Bright interference fringes appear in the recording plane where the quantity

$$\Delta l(x, y)/\lambda = 0, \pm 1, \pm 2, \ldots. \tag{3.33}$$

Equation (3.33) is the equation of the fringes in the recording plane (x, y). The integer numbers on the right-hand side of Eq. (3.33) are the orders of the bright fringes. In between each two bright fringes is a dark fringe for which $\Delta l/\lambda = \pm\frac{1}{2}, \pm\frac{3}{2}, \ldots$. The fringe pattern ("interferogram") is the information that can be measured, and from Eq. (3.32) we see that this information, expressed by $\Delta l(x, y)$, is available in two-dimensional form; whereas, the quantity to be determined, the refractive index n or the fluid density ρ, is a function of the three coordinates x, y, z. In Section 3.3.5 it will be discussed how this two-dimensional information can be converted into values of the three-dimensional density field $\rho(x, y, z)$. This problem does not exist for a plane density field $\rho(x, y)$ in which the density does not change in the z direction.

A further problem incorporated in Eq. (3.32) is the presence of two unknowns, namely the values of n at two different positions. The solution to this problem, the necessary reduction of the number of unknowns from two to one, is achieved by an appropriate design of the "interferometer unit," our previous black box. The separation d between the two conjugate rays cannot be arbitrary as assumed so far, and for solving our problem, two extreme situations are possible, which correspond to two classes of optical interferometers. If D designates the diameter of the field of view, this classification of two-beam interferometers can be described by the ratio d/D:

1. With $d/D \geqq 1$, one ray of each pair of conjugate rays propagates outside of the test field and remains undisturbed. This is the class of *reference beam interferometers.*
2. For a ratio $d/D \ll 1$, both rays traverse the test field, where they are separated or sheared by the small distance d. Such systems are often called *shearing interferometers.*

Since one ray of each pair passes along a constant refractive index n_∞, the equation of the fringes for the reference interferometer is

$$\frac{1}{\lambda} \int_{\zeta_1}^{\zeta_2} \{n(x, y, z) - n_\infty\}\, dz = 0, \pm 1, \pm 2, \ldots. \tag{3.34}$$

It is seen that for a plane density distribution with $n = n(x, y)$ and with $b = \int_{\zeta_1}^{\zeta_2} dz$ being the width of the test field, the fringes

$$(b/\lambda)\{n(x, y) - n_\infty\} = 0, \pm 1, \pm 2, \ldots \tag{3.34a}$$

are curves of constant refractive index or constant fluid density.

The two integrals in Eq. (3.32) can be developed in Taylor series around the value of $n(x, y, z)$. Since d is a small quantity in the case of a shearing interferometer, it is permitted to take into account only the linear terms (the quadratic terms vanish). The result, written in form of the equation of fringes for a shearing interferometer, is

$$\frac{d}{\lambda} \int_{\zeta_1}^{\zeta_2} \frac{\partial}{\partial y} n(x, y, z)\, dz = 0, \pm 1, \pm 2, \ldots \tag{3.35}$$

We realize that the shearing interferometer is sensitive to changes of the refractive index gradient. This also led to the name "schlieren interferometer," which is occasionally used for this class of instruments. Indeed, the form of Eq. (3.35) is similar to Eq. (3.24), which describes the contrast in the recording plane of the schlieren system. The derivative $\partial n/\partial x$ is obtained if the optical arrangement (Fig. 3.28) is rotated by 90° around the optical axis. For a two-dimensional density field, the fringes as described by

$$\frac{bd}{\lambda}\frac{\partial}{\partial y} n(x, y) = 0, \pm 1, \pm 2, \ldots \tag{3.36}$$

are curves of constant derivative of the refractive index or density.

Equations (3.34) and (3.35), respectively, apply to an alignment of the interferometer for which a uniform test field (with n = const. or $\partial n/\partial y$ = const., respectively) appears uniformly illuminated, that is, no fringe is seen in the field of view. This case is called the "infinite fringe width" alignment. Both interferometers can be aligned in a different way, so that a system of equidistant, parallel interference fringes appears for a uniform test field. A density variation in the flow will distort this regular fringe pattern, and the deviation or shift of a fringe from its undisturbed position is a measure for the density disturbance. Such an alignment is called the "finite fringe width alignment." A fringe shift by one fringe width is equivalent to a difference in optical path length by one wavelength. It follows that a fringe shift ΔS in a point (x, y) measured in terms of fringe width S of the undisturbed, equidistant pattern, is

$$\frac{\Delta S(x, y)}{S} = \frac{1}{\lambda}\int_{\zeta_1}^{\zeta_2} \{n(x, y, z) - n_\infty\}\, dz \tag{3.37a}$$

for the reference beam interferometer, and

$$\frac{\Delta S(x, y)}{S} = \frac{d}{\lambda}\int_{\zeta_1}^{\zeta_2} \frac{\partial}{\partial y} n(x, y, z)\, dz \tag{3.37b}$$

for the shearing interferometer. The curves connecting the points in which the disturbed fringes intersect with the undisturbed fringes of the finite fringe width case are identical with the interference fringes of the infinite fringe width alignment.

In the following two sections it will be discussed how the two sorts of interferometers can be realized technically.

3.3.2. Reference Beam Interferometers

A reference interferometer needs a beam splitter that separates the coherent light from the light source into two beams, the test beam and the reference beam. We will see that, in a Mach–Zehnder interferometer, this separation is accomplished by means of a plane splitter plate; whereas, in a holographic interferometer the splitting of the two beams is converted into a separation in time of their existence. These two instruments are by far the reference beam interferometers most often used for flow visualization. Other reference beam interferometers, though being popular in other fields of application, do not play a substantial role here. Even the well-known Michelson interferometer has found a very limited number of applications (e.g., Gille, 1967).

In some cases it is of interest to convert a schlieren system into an interferometer. This can be accomplished by using two diffraction gratings as the beam splitter and the recombining element (interferometer unit), respectively. In such a "grating interferometer" (Kraushaar, 1950; Sterret *et al.*, 1965; Maddox and Binder, 1969), a grating of about 1000 lines/cm is placed in the front of the first schlieren lens; for example, at the position of the light source in Fig. 3.11, while the light source, now placed to the left of this position, must be focused into this point. Two diffraction orders, usually the zeroth and first order, propagate as parallel beams to the second lens (schlieren head) where they are refocused onto the second grating. The test field is placed in one of the parallel beams, which serves as the test beam. The diffraction process at the second grating provides that a diffraction order of the test beam and another diffracted order of the reference beam overlap, thus producing a reference interferogram. Besides the ease of using an available schlieren system, this grating interferometer has a number of technical disadvantages; and it will be seen later that it is more convenient to convert a schlieren system into a shearing interferometer (Section 3.3.3) or to generate reference interferograms by manipulating the light distribution in the focal plane of the schlieren head (Section 3.4).

3.3.2.1. Mach–Zehnder Interferometer

A reference beam interferometer designed by Jamin in 1856 (Fig. 3.29) had been used by Mach and von Weltrubsky (1878) to study gas dynamic phenomena. Ernst Mach recognized that for better applicability of the method it was desirable to have an instrument with the test and the reference beam more widely separated. The practical realization of this idea was developed independently by Zehnder (1891) and Ernst Mach's son Ludwig Mach (1892). Since the days of its innovation the instrument and the methods for evaluating interferograms have been improved continuously, as reviewed, e.g., by Schardin (1933), Kinder (1946), Winckler (1948), and Ladenburg and Bershader (1954).

The Mach–Zehnder interferometer (MZI) combines a wide separation

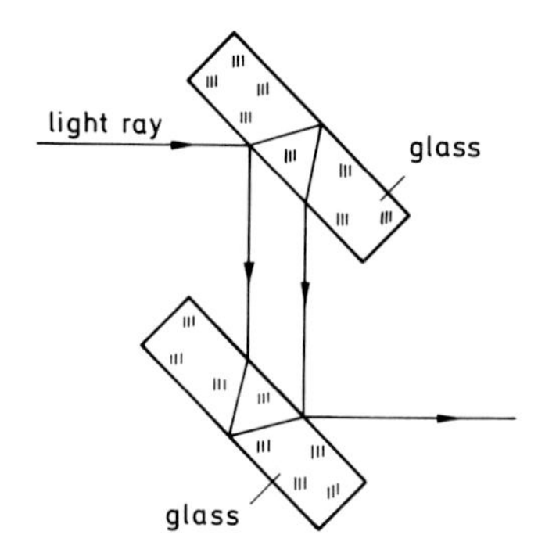

Fig. 3.29 Two-beam interferometer of Jamin.

of test beam and reference beam (i.e., $d/D > 1$) with a relatively large diameter D of the test beam and the field of view. In a basic arrangement (Fig. 3.30), light from a point source is made parallel with the lens or spherical mirror L_1. The essential components of the interferometer are the plane, fully reflecting mirrors M_1 and M_2, and the plane, semireflecting mirrors (beam splitters) M_1' and M_2'. These four mirrors are arranged here to form a rectangle. The test section of the flow facility with its two glass windows is brought into the path of the test beam, while two identical glass plates are inserted into the path of the reference beam in order to compensate for the large optical path length travelled by the test beam in the glass. The role of M_2' is that of the aforementioned interferometer unit (Fig. 3.28), recombining the initially separated test and reference beam so that they coincide and interfere with one another. The camera film is focused onto a plane in the test section.

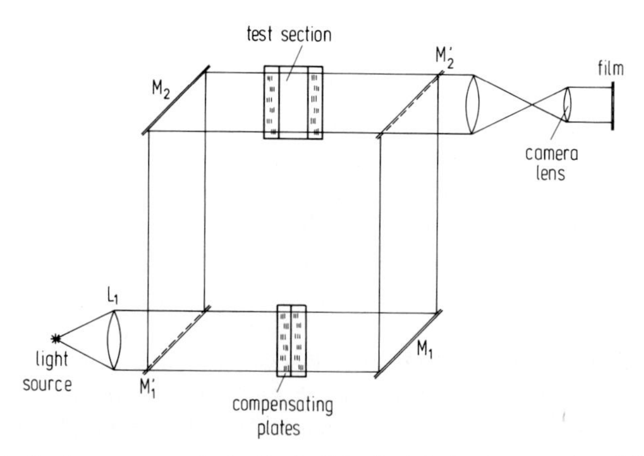

Fig. 3.30 Basic arrangement of a Mach–Zehnder interferometer, where M_1, M_2 are the mirrors, and M_1', M_2' the beam splitters.

Each of the four mirrors is required to permit rotation around a horizontal and a vertical axis for the basic adjustment of the interferometer. If (in the arrangement shown in Fig. 3.30) all mirrors are inclined at exactly 45° to the incident parallel light beam, and if all optical components are assumed to be of perfect quality, the light arriving in the recording plane is in phase, and no fringes appear in this basic position of the interferometer. This is the "infinite fringe width" alignment. The finite fringe width alignment can be realized by tilting one plate (e.g., M_2') by a small angle ε; the rays r_1 and r_2 intersecting in P (Fig. 3.31) exhibit a phase difference owing to their difference in path length

$$\Delta l = \mathbf{OP} - \mathbf{QP} = b\,\frac{1 - \cos(2\varepsilon)}{\sin(2\varepsilon)} = b \tan \varepsilon \cong b\varepsilon. \qquad (3.38)$$

Tilting M_2' is equivalent to inserting a transparent wedge of angle ε into the path of one light beam. An observer looking from the recording plane into

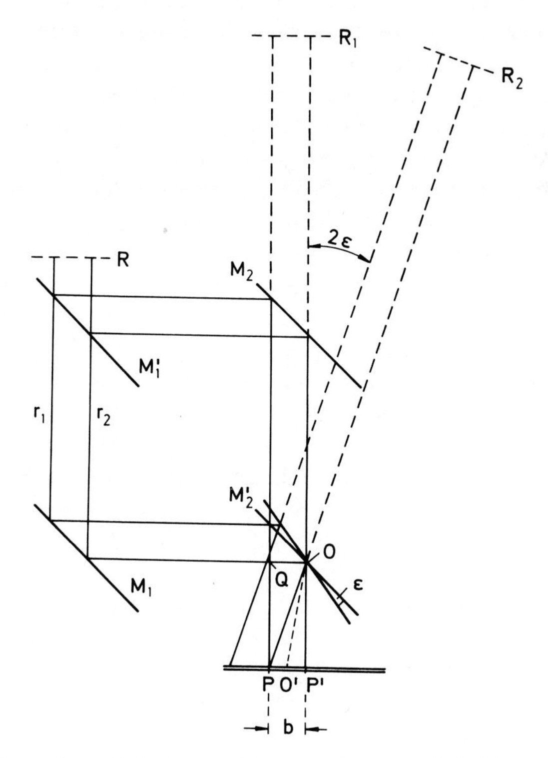

Fig. 3.31 Formation of interference fringes on the screen PO′P′ by inclining the beam splitter M_2' by an angle ε.

the interferometer would see two light sources separated by an angle 2ε, and without knowing of the mirror system in the interferometer, he assumes the light beams to be straight, as indicated in Fig. 3.31. In the linear representation of the MZI shown in the figure, the planes R_1 and R_2 correspond to the plane R in the real system, and they coincide, if the interferometer is in its basic position ($\varepsilon = 0$). In the point O and on the bisector of the angle 2ε the interfering waves have zero phase difference, and the fringe of zeroth order is located in O' (in the recording plane $\mathbf{P}O'\mathbf{P}'$). If $\Delta l = N\lambda$, with λ being the wavelength, there will be N fringes between O' and P (fringe direction is normal to the plane of the figure), and

$$N\lambda = \varepsilon b,$$

while the fringe width S is given by

$$S = b/2N = \lambda/2\varepsilon. \tag{3.39}$$

The result is equivalent to the treatment of Young's fringes in connection with speckle photography (Fig. 2.27). The fringe direction can be altered by rotating the tilted mirror M_2'.

The interference in the point P in Fig. 3.31 is established by two rays, which originate from two different points of the plane R. This requires all points of R to be coherent, which is not necessarily the case. It is therefore more adequate that the interference be produced by the two components ("conjugate rays") of one single ray separated at the beam splitter M_1'. This situation can be realized, e.g., by tilting both M_1' and M_2'. One may then distinguish between two possible cases sketched in Fig. 3.32. Tilt angles of M_1' and M_2' are ε_1 and ε_2, respectively. In the first case (left of Fig. 3.32) the interfering rays intersect after having left the interferometer. The locus of intersection of conjugate rays is the location of interference, and the fringes in this case are real (i.e., they are formed in the recording plane). In the second case, the two plates are tilted in a way causing the conjugate rays to intersect within the interferometer (i.e., before they reach M_2'). The fringes are now called virtual. It is possible to focus this position of interference onto the recording plane where a system of sharp fringes will be observed.

The problem in the latter case is to have both the fringes and the object in focus. A further problem is to reduce the number of plates that have to be tilted to adjust the interference fringes. An instrument with single plate control that fulfills these two requirements has been described by Kinder (1946). The four mirrors are arranged in form of a rectangle, with M_2 being twice as far from M_2' as M_1, and the test section is inserted midway between M_2 and M_2' (see the linear representation in Fig. 3.33). While M_1',

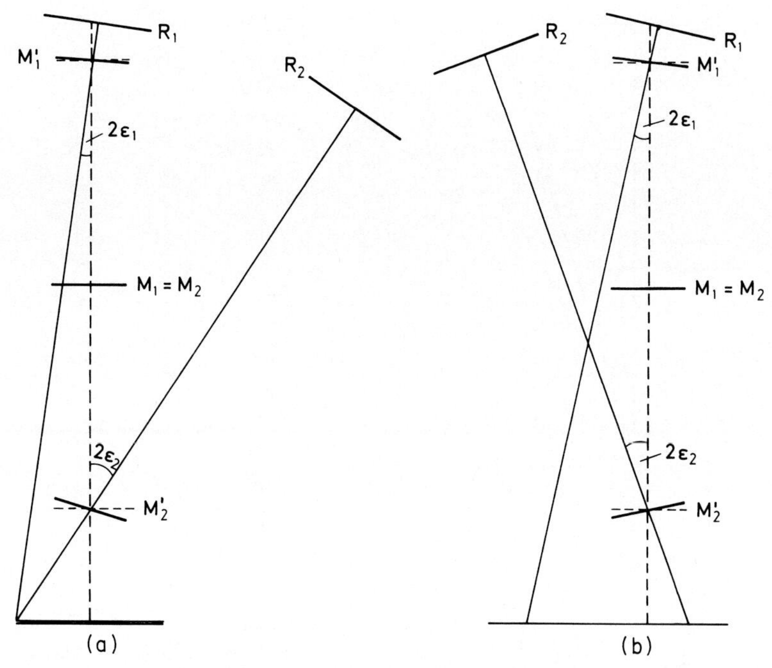

Fig. 3.32 Linear representation of the MZI with inclined beam splitters M_1' and M_2'. Formation of (a) real fringes and (b) virtual fringes.

M_2, and M_2' remain in the basic position under 45° of inclination, only M_1 has to be rotated for adjusting the fringes. The fringes are located in M_1, or in a position very close to M_1, and they appear in focus, if the test section is imaged onto the recording plane.

High-precision mechanical performance and a high quality of the components are the essential requirements for the construction of a Mach–Zehnder interferometer. Mechanical and optical tolerances of surface flatness, surface parallelism, translational and rotational displacement, control of tilt, are in the order of one tenth of a wavelength. This makes the instrument expensive, and the costs grow rapidly with increasing diameter of the plates (i.e., with the desired size of the field of view). A test beam diameter of 200 mm is possible. The plates are mounted on a U-shaped steel frame forming three sides of a rectangle or a parallelogram. The fourth side remains open for inserting the test object. The required mechanical stability of the frame limits the size of the rectangle, and thereby the possible separation of test and reference beam. Instruments are commercially available. The one shown in Fig. 3.34 is designed to give space for a test section of an outer diameter of 1 m.

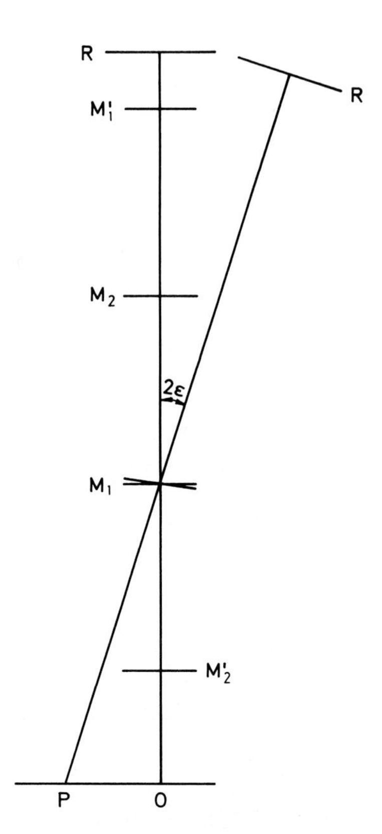

Fig. 3.33 Linear representation of the MZI with single-plate control.

The best quality control of the instrument is to check the parallelism of the interference fringes in the finite fringe width mode (this criterion applies to any interferometer). A deviation of not more than one tenth of a fringe width is desirable. This results in a uniformly illuminated field of view in the infinite fringe width mode, or the appearance of only one color in this mode when a polychromatic (white) light source is used. When monochromatic light sources (lasers) were not available, the basic alignment of a MZI was a tedious procedure, sometimes requiring several days of work. With the use of coherent laser light sources this alignment has become a relatively easy job.

The precision requirements also apply to the test section windows (if there are any), their surface finishing and optical homogeneity of the glass. Optically inhomogeneous glass produces an additional distortion of the interference fringes, or in the worst case it totally suppresses interference between the two beams. Such an extreme situation is depicted in Fig. 3.35 where a hot light bulb is placed in the test beam. The Mach–

Fig. 3.34 Mach–Zehnder interferometer manufactured by Zeiss, Oberkochen, Germany.

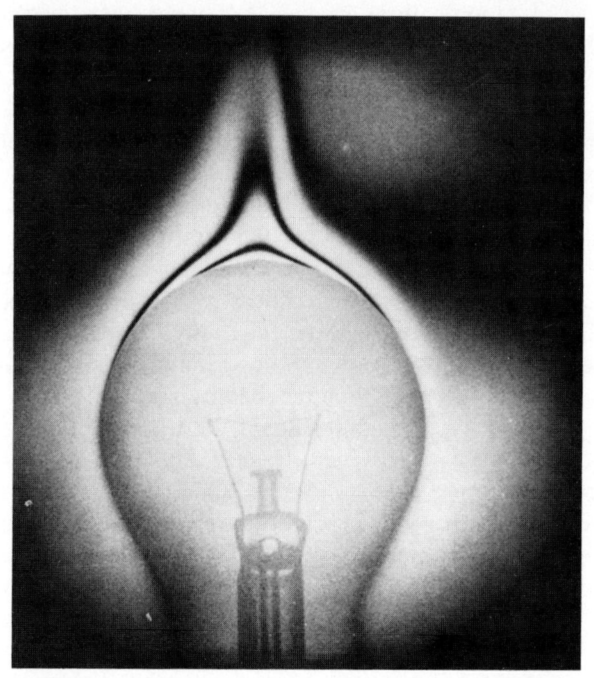

Fig. 3.35 Mach–Zehnder interferogram showing the density field in the plume rising from a hot light bulb. The poor quality of the glass does not allow for optical interference for the light transmitted through the bulb. (From A. Metzger and W. Merzkirch, Ruhr-Universität Bochum.)

Zehnder interferogram in the finite fringe width mode shows fringes resulting from the plume of warm air above the bulb. Figure 3.35 is at the same time an example for the interferometric visualization of a free convective flow. The glass of the bulb is so poor that the interferogram does not include information on the temperature field in the interior of the bulb. There is, at least, a principal possibility how to compensate for such impurities in the glass: The MZI has two "exits;" only one is used in Fig. 3.30, namely to the right of the plate M_2'. A second exit appears if the vertical beam from M_1 to M_2' is extended in vertical direction. The fringe pattern recorded in the second exit is complementary to that observed in the first exit, that is, dark fringes appear at positions where bright fringes exist in the interferogram of the first exit, and vice versa. The compensation procedure would require to take an interferogram without flow (but with the impure test section windows in place) in one exit, and to superimpose this record to the interferogram taken at the second exit in the presence of the flow. The fringe pattern caused by the disturbances from the glass will then be cancelled, and only the pattern produced by the flow will remain.

A number of modifications of the MZI have been discussed in the literature. Howes (1984) describes an arrangement, which overcomes the limitations in the size of the field of view. A schlieren system of large aperture is followed by a relatively small MZI-like setup (Fig. 3.36). The reference beam of the MZI is fed through a spatial filter where all the phase information from the test object is removed. Large object fields are allowed in the schlieren system, and the interferometric arrangement can be kept small. Yokozeki and Mihara (1979) show how moiré patterns can be produced by superimposing two Mach–Zehnder interferograms taken in the finite fringe width mode. The linear representation of the interferometer in Figs. 3.31–3.33 is a means for easily comparing the MZI with other reference beam interferometers, for example, an interferometer described by Xia (1985) in which the test beam, like in the Michelson interferometer, passes twice through the flow field, thereby doubling the sensitivity.

The number of applications of Mach–Zehnder interferometry to compressible aerodynamics, both in wind tunnels and in shock tubes, is immense and we refrain from giving specific references. Figure 3.37 gives an example for a finite and an infinite fringe width interferogram taken in a supersonic wind tunnel. Buxmann (1970) reports on the application of the MZI to the flow over compressor blades; Anderson *et al.* (1977) discuss the possibility of obtaining data from a turbulent flow field with fluctuating density. Also numerous are the applications to convective heat transfer problems, both in gases (e.g., Pera and Gebhart, 1975) and in liquids

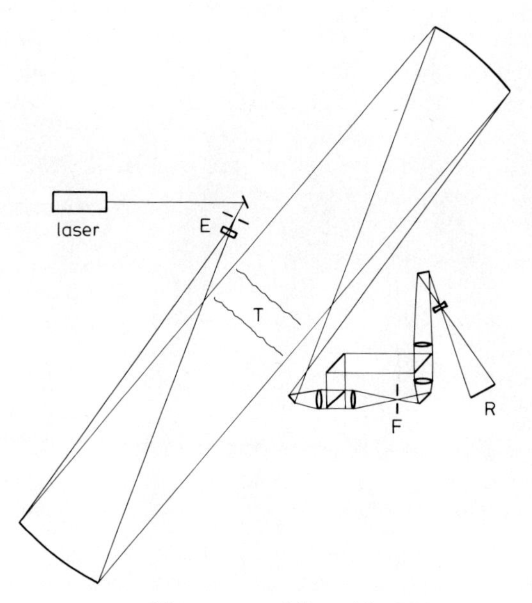

Fig. 3.36 Large aperture schlieren setup followed by MZI arrangement in which the "reference beam" is fed through a spatial (pinhole) filter (F). E designates the beam expander; T, test field; R, recording plane. (After Howes, 1984.)

(Bathelt and Viskanta, 1980). Specific error sources exist for these applications where the aim is to measure either the fluid temperature or the heat transfer rate in terms of the temperature gradient. Flack (1978a) examines the errors arising from a misalignment of the plane walls in the test field where the heat transfer is to be measured; Koster (1983) discusses the errors caused by a secondary temperature field developing in the test section windows. Stevenson *et al.* (1983) visualized internal gravity waves formed in stratified saltwater.

3.3.2.2. Holographic Interferometry

Holography has literally opened a new dimension for two-beam interferometry: Instead of the spatial separation of reference and test beam, the two beams can now be separated in the time of their existence. The key is that it is possible to store on one holographic plate the information of two (or more) light waves and to simultaneously release these originally separate informations. The reconstructed waves overlap and can interfere if the conditions of coherence are fulfilled. In a holographic interferometer two consecutive holographic exposures are taken through the field of interest, usually the first exposure without flow, the second in the presence of the flow field with varying fluid density. The first exposure is

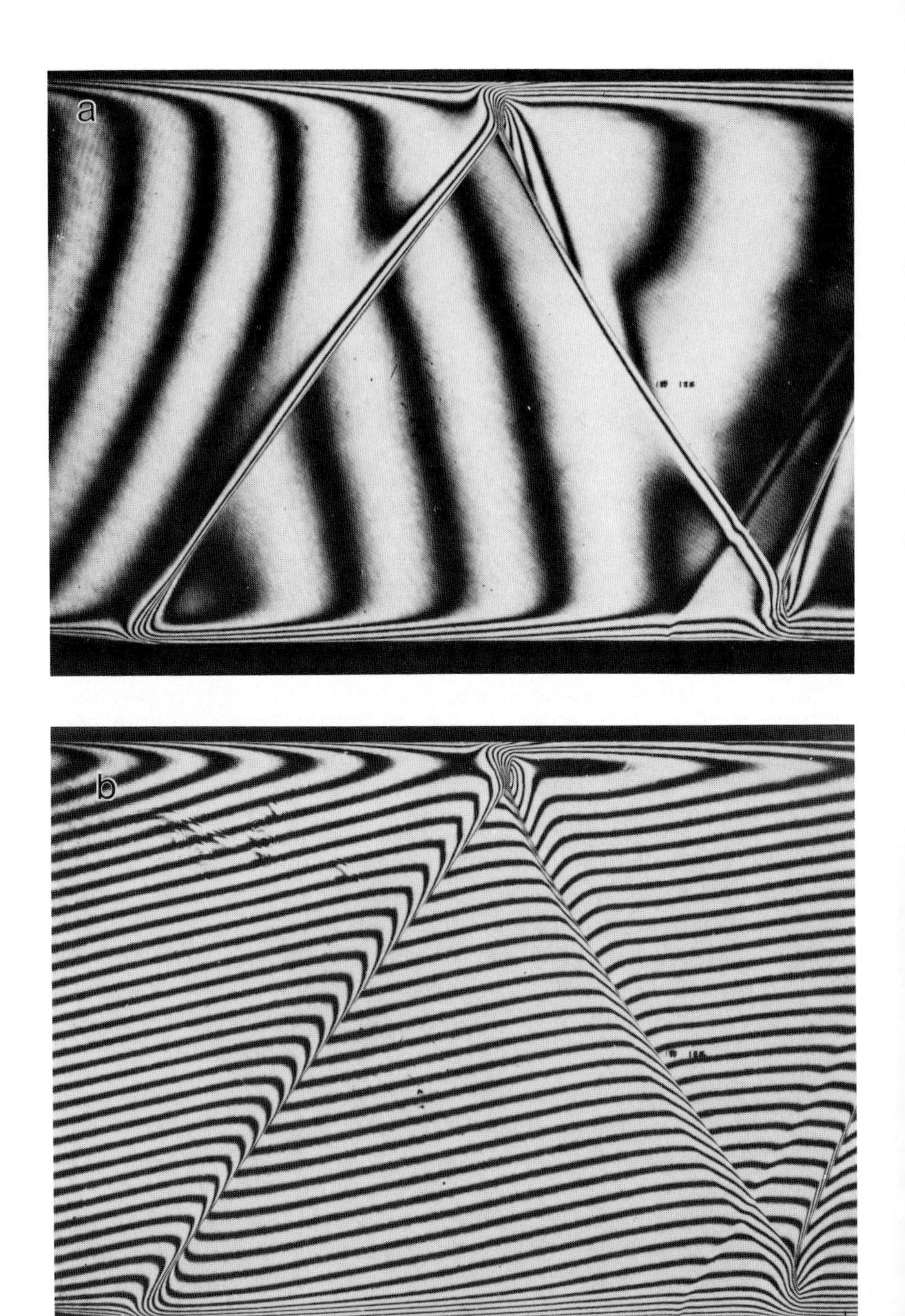

Fig. 3.37 Reference beam interferogram of the flow in a transonic wind tunnel at (a) infinite fringe width, and (b) finite fringe width alignment. (From Délery *et al.*, 1977.)

equivalent to the reference beam of a Mach–Zehnder interferometer, the second exposure carries the information of the test beam. Upon simultaneous reconstruction of the two holographic recordings, one produces a wave pattern that is the superposition of the two light waves frozen in the two separate exposures. Thus, the same interference pattern is obtained as observed in a Mach–Zehnder interferometer, where the two light waves, representing the test and the reference beam, exist "in real time."

Holographic interferometry was introduced to flow visualization and the measurement of density fields by Heflinger *et al.* (1966) and Tanner (1966a). When it became evident that holographic interferometry has a number of essential advantages over classical Mach–Zehnder interferometry, the new method was soon picked up by a number of laboratories (e.g., Philbert and Surget, 1968; Smigielski and Royer, 1968; Reinheimer *et al.*, 1970), and the number of applications and systems developed grew rapidly. Some laboratories produced technical guidelines for the use of this novel technique. The book of Vest (1979) is a detailed discussion of the principles and a review of the state of the art.

In the basic setup of a holographic interferometer (Fig. 3.38) a laser beam is separated by a beam splitter into two beams. One of them, after being expanded, traverses the test field as a collimated beam and then strikes the holographic plate (film) in the recording plane. In holographic interferometry, this beam usually is named the test beam; however, in order to avoid confusion with the test beam of a reference beam interferometer, we designate this beam the signal beam. The second beam is fed outside of the test field, it remains undisturbed and, after expansion by a

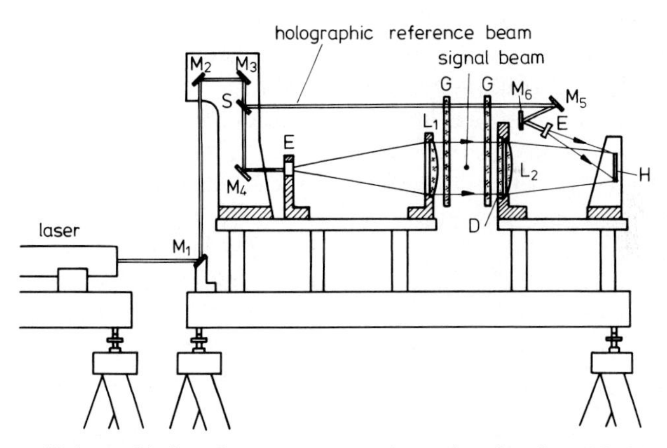

Fig. 3.38 Holographic interferometer mounted on a bench, where M_1–M_6 are the reflecting mirrors; S, beam splitter; E, beam expander; L_1 and L_2, lenses; G, glass windows of test section; D, diffuser (ground glass); and H, holographic plate. (From Délery *et al.*, 1977.)

second beam expander, it also strikes the holographic plate. We will call this second beam the holographic reference beam, and again, this beam should not be confused with the reference beam of a reference beam interferometer. The holographic plate is a photographic material with a resolution finer than normal photographic film. The setup in Fig. 3.38 is shown as a vertical arrangement on an optical bench. A different and frequently used setup is horizontal and arranged on a holographic table, a heavy granite plate that provides to the optical arrangement mechanical stability.

By exposing the holographic plate with both the signal beam and the holographic reference beam and subsequent developing, one produces a hologram of the test field. If the developed holographic plate ("hologram") is replaced in its position in the setup of Fig. 3.38 and illuminated there with the holographic reference beam only, the signal beam is reconstructed (i.e., the test field becomes visible even if it has been removed from the setup). We refrain here from explaining the details of the holographic principle, and the reader is referred to the respective literature (e.g., Vest, 1979). Making the test field visible in the described way does not mean, that the flow is made visible. The reconstructed signal beam carries the same information, particularly on the distribution of the optical phase, as the test beam of a Mach–Zehnder interferometer, and what is needed for an interferometric visualization is a reference beam analogous to that in the MZI. This reference information is created by taking another holographic exposure of the test facility, but without flow. The two exposures can be recorded on the same holographic plate ("double exposure"). Usually, one first takes the reference exposure without flow, and then the test exposure with flow. The double-exposed plate is developed, replaced in the holographic reference beam for reconstruction, and the two signal beams (reference and test beam) are reconstructed simultaneously and can interfere with one another. The pattern that becomes visible in this reconstruction process is an infinite fringe width interferogram carrying the same information on the density field as a respective Mach–Zehnder interferogram.

There are a number of fundamental differences with respect to Mach–Zehnder interferometry. Since the geometrical path of the signal beams in the two exposures is identical, the optical paths of the two signal beams differ only due to the difference in fluid density between the flow and the no-flow situation. Optical disturbances or impurities in the glass of the test section windows or in any other optical component of the setup are cancelled out, and a holographic interferogram can be taken through glass windows of relatively poor quality. This is demonstrated in Fig. 3.39 where the temperature field inside the hot bulb is now visible, in contrast

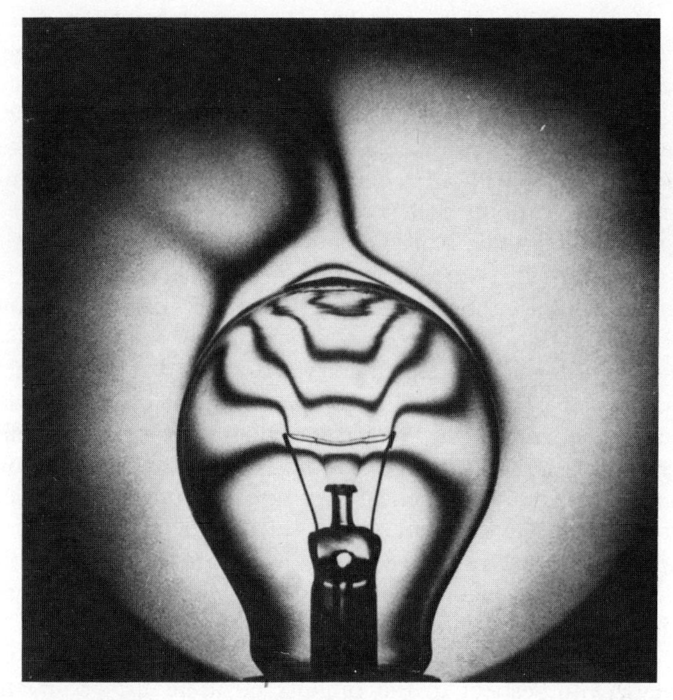

Fig. 3.39 Holographic interferogram showing the density field inside and outside of a hot light bulb. Distortion of the light due to the poor quality of the glass is compensated for in the holographic interferometer, see also Fig. 3.35. (From A. Metzger and W. Merzkirch, Ruhr–Universität Bochum.)

to the Mach–Zehnder interferogram (Fig. 3.35) that could not be recorded through the glass of the bulb. Another difference is that, with this double exposure on one plate, it is not possible to observe the interference pattern "in real time" (i.e., during the experiment). The frozen pattern becomes visible only after development and during reconstruction. A further question is how to produce a finite fringe width pattern. Solutions to these two problems will be discussed next.

Real time interferometry is particularly desirable if the flow pattern changes with time. The pattern could then be observed directly or recorded with a movie camera. Real time observation requires that the reference beam is existent during the time of the experiment. This can be achieved by taking the reference exposure (without flow) first, developing this exposure and replacing the plate in its original position in the holographic setup. During the experiment, both the holographic reference beam and the signal beam illuminate the no-flow hologram, and the reconstructed reference beam interferes with the existing test beam. The fringe pattern, which becomes visible if one looks from behind through the holographic plate, can be recorded with a cine camera (e.g., Surget and Chatriot, 1969; Aung and O'Reagan, 1971; Dullforce and Faw, 1979). The developed plate of the reference exposure has to be replaced at exactly the original position (with an accuracy of a fraction of a wavelength), if the infinite fringe width mode has to be matched. This requirement led to the design of holographic plate holders with high resolution positioning control, and alternatively to plate holders in which the developing process can be done onsite without removing the plate (Achia and Thompson, 1972).

From Mach–Zehnder interferometry we know that, for generating a finite width fringe pattern, one has to provide a small angle between test and reference beam. In the holographic double-exposure technique, it would be required to rotate the plate after the first (reference) exposure by a small angle, before taking the test exposure. But this is impractical, and it is more desirable to generate the finite fringe width mode during the reconstruction process. If the reference hologram for real time observations is not replaced precisely in its original position, fringes of finite width will appear in the undisturbed field of view, the width and orientation of the fringes depending on the amount of displacement and rotation of the plate with respect to the original position. Fringe spacing and orientation can be controlled arbitrarily with the dual-plate holder: Reference and test hologram are taken on two separate plates, which, after being developed, can be rotated and sheared with respect to each other in this holder during the simultaneous reconstruction (Hannah and Havener, 1975; Dunagan *et al.*, 1985). Burner and Goad (1981) take the test exposure first, process this hologram, and adjust the reference beam in real time during reconstruction for generating the appropriate finite fringe width pattern.

Another solution for realizing the finite fringe width mode is a holographic setup using the two separate holographic reference beams, one for the reference exposure and a second of different geometry for the test exposure (Surget, 1974; Dändliker *et al.*, 1976; Trolinger, 1979; Farrell *et al.*, 1982). The two holographic reference beams strike the same holographic plate (Fig. 3.40), each at the respective exposure, and a double exposure on one single plate can be taken. For producing an interferometric pattern in the desired mode, the hologram is illuminated by the two holographic reference beams simultaneously for reconstruction. Controlling the angle between the two beams allows for selecting the appropriate fringe pattern. Upon illumination with only one of the holographic reference beams, each object state, with or without flow, can be reconstructed

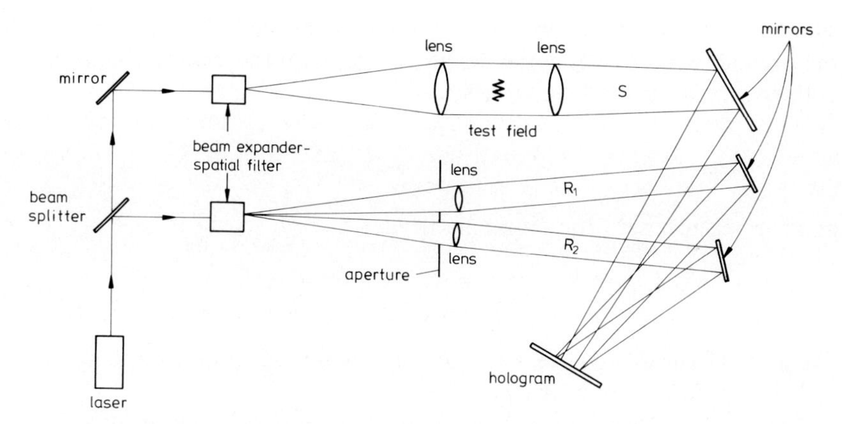

Fig. 3.40 Holographic interferometer with two holographic reference beams, R_1 and R_2; S designates the signal beam. (From Trolinger, 1979. Copyright, Optical Society of America.)

individually. This is of interest if one wants to investigate the reconstructed test beam in a schlieren system or in another interferometer (see also Fig. 3.18). An angle between the reconstructed test and reference beam can also be generated, if the two individual exposures are recorded with different wavelengths, but reconstructed with only one wavelength (Walklate, 1981).

Ultimately, the aspect that holography provides three-dimensional information should be discussed. A certain kind of spatial resolution is obtained if one inserts a ground glass in the path of the signal beam immediately in front of or behind the test field (the diffuser D in Fig. 3.38). Since this diffuser scatters the light in various directions, we now have a continuum of beam directions traversing the test field, and the reconstructed scene may be examined under different viewing angles. In the following we will investigate how the observable fringe pattern is dependent on the viewing direction. When one focuses the eye or a camera onto an object plane in the reconstructed scene light, one does not necessarily observe any interference fringes: A variety of light rays of different direction passes through every point of the object plane, and all these rays are collected in the respective image point. Generally, the optical paths along these rays are different, and the associated waves arrive in the image point with random optical phase. The phase differences with respect to the reference wave might average out and, consequently, no fringes are seen.

In order to generate interference fringes it is necessary that the optical path lengths l_{opt} of all rays passing with different direction through the same object point be independent of the viewing angle α: $(dl_{opt}/d\alpha) = 0$. If the optical axis ($\alpha = 0$) coincides, as usual, with the z-direction, one has with the denotation of Section 3.1

$$l_{opt} = \int_{\zeta_1}^{\zeta_2} n(x,y,z) \frac{dz}{\cos \alpha}, \tag{3.40}$$

and the requirement for obtaining fringes is

$$\frac{dl_{opt}}{d\alpha} = \int_{\zeta_1}^{\zeta_2} \left\{ n \tan \alpha + \frac{\partial n}{\partial x}\frac{dx}{d\alpha} + \frac{\partial n}{\partial y}\frac{dy}{d\alpha} \right\} \frac{dz}{\cos \alpha} = 0. \tag{3.41}$$

Equation (3.41) designates a surface onto which the recording system has to be focused to obtain fringes.

Figure 3.41 shows an arrangement, which enables one to observe an interference pattern generated by a parallel beam of light traversing the test field in a direction α_1. For simplicity, the object itself is shown here and not the reconstruction setup (in which a virtual image of the scene is produced). The direction α_1 is separated from the rest of the light by means of a small aperture in the focal plane of the lens Le. Other light

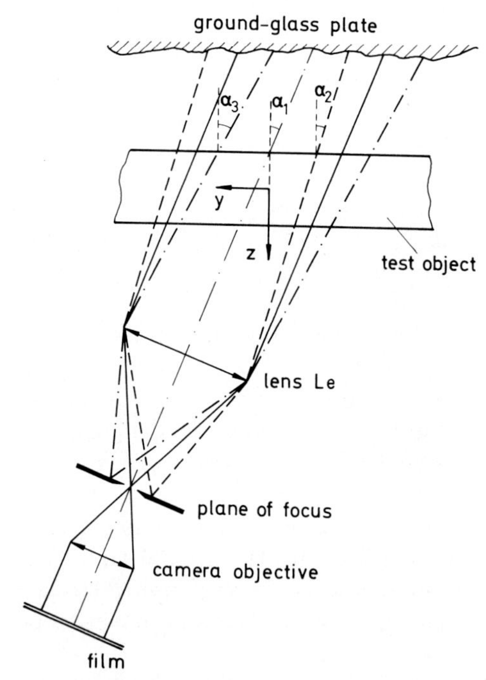

Fig. 3.41 Observation of the interference pattern produced by a parallel light beam passing at a direction α_1 through the test field. Light directions α_2 and α_3 from the diffuser are blocked off in the focal plane of lens Le.

directions (e.g., α_2 and α_3 in Fig. 3.41) are blocked off. Since the diffuser produces a continuum of light directions, the aperture has to be small in order to select a narrow range of directions α. The influence of diffractive noise is thereby increased, and the interference fringes loose their sharpness and clarity (Tanner, 1968). Vest and Sweeney (1970) have demonstrated that this noise problem can be reduced if one replaces the ground glass by a diffraction grating, thereby limiting the possible viewing directions to a discrete number. The total coverage of possible viewing angles is determined by the characteristics of the ground glass or the grating, respectively. Fig. 3.42 includes three infinite fringe width interferograms

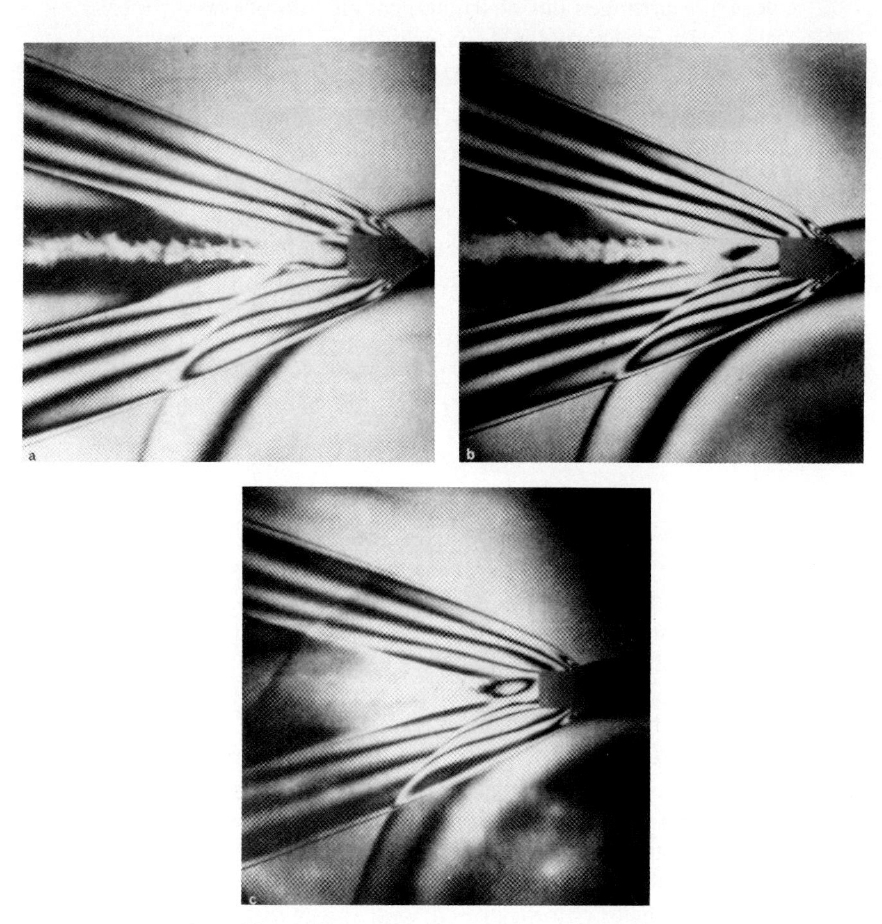

Fig. 3.42 Infinite fringe width holographic interferogram showing the interaction of a supersonic projectile with a blast wave: (a) $-20°$ view, (b) $0°$ view, and (c) $+20°$ view. Note the different relative positions of the blast wave with respect to the projectile. (From Witte and Wuerker, 1970.)

taken under different viewing angles (−20°, 0°, +20°) of a three-dimensional density field originating from the interaction of a supersonic projectile with a blast wave.

A particular aspect of holographic interferometry is that it becomes possible to direct the signal beam twice or several times through the test section, thereby increasing the effective width of the test field and the sensitivity toward density changes (Clark, 1978; Kaiser, 1984). In summarizing, the following fundamental differences to Mach–Zehnder interferometry can be identified:

(1) compensation of impurities in test section windows;
(2) versatile arrangement of fringe spacing and orientation;
(3) observation of the object under different viewing angles; and
(4) possibility of multipass interferometry.

The question whether the setup for either one of the techniques is less expensive is hard to answer; different claims have been made on this subject. Besides the references already given, a great number of applications in various fields has been reported, a few of which will be referred to as characteristic examples: high-speed aerodynamics (Lee *et al.*, 1984), turbulence diagnostics (Trolinger and Simpson, 1979), turbines and compressors (Bryanston-Cross *et al.*, 1981; Decker, 1981), heat and mass transfer (Masliyah and Nguyen, 1979; Mayinger and Steinberner, 1979; Faw and Dullforce, 1981; Hatfield and Edwards, 1981), stratified flow (Debler and Vest, 1977), underwater shock waves (Takayama and Onodera, 1985).

3.3.3. Shearing Interferometers

3.3.3.1. Shearing Elements

An optical element is needed for providing the shear between the two beams, which both traverse the test field and then interfere with one another. In Fig. 3.28 it is assumed that a shear of amount d is generated in the y direction; whereas, no shear exists in the x direction. Instruments that provide such a constant shear in one direction, normal to the optical axis, are called lateral shearing interferometers. It is also possible to generate the shear in radial direction with respect to the optical axis, and the respective instruments are named radial shearing interferometers. A great variety of shearing elements, playing the role of the interferometer unit in Fig. 3.28, has been described, and the respective literature has been reviewed occasionally (Tanner, 1966b; Bryngdahl and Lee, 1974; Merzkirch, 1974). In many cases, a schlieren system can be easily converted into a shearing interferometer, if the respective shearing element is used in place of the conventional knife edge. This allows one to investigate by interferometry extended test fields of large diameter (Popovich and Weinberg, 1983).

Among the many possible shearing elements the Wollaston prism has found the most frequent application in flow visualization (see the Section 3.3.3.2). The simplest elements are a flat glass plate (Kelley and Hargreaves, 1970) or a flat glass wedge (Sandhu and Weinberg, 1972), with the front surface partially reflecting and the rear surface totally reflecting. Light rays reflecting from the front surface interfere with rays reflecting from the rear surface; the thickness of the plate, the angle of incidence, or the wedge angle, respectively, determine the amount of the generated shear, d (Fig. 3.43). An optical setup with a wedge as the shearing element is shown in Fig. 3.44. This setup demonstrates the feasibility of studying a large flow field with an interferometer of relatively small aperture.

3.3.3.2. *Wollaston Prism Interferometer*

This shearing interferometer was originally developed for use in interference microscopes by Francon (1952) and Nomarski (1956). Since 1957, this principle has been applied by various authors to visualize compressible flows and to investigate quantitatively certain gas dynamic problems [e.g., by Chevalerias *et al.* (1957), Gontier (1957), Philbert (1958), Merzkirch (1965), Oertel (1967), and Smeets (1968)]. We will first discuss the role of the Wollaston prism in the optical setup and then investigate the conditions under which interference fringes are formed with this system.

The prism described by Wollaston in 1820 is a double prism made of quartz or calcite. Its purpose is to separate an incident light ray into two

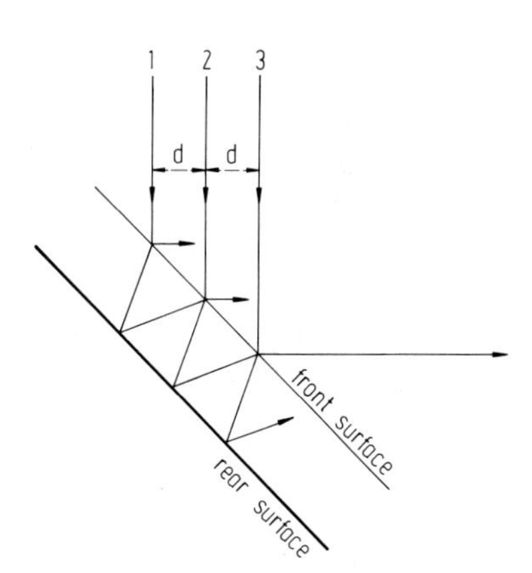

Fig. 3.43 Second surface mirror used as a shearing element (interferometer unit).

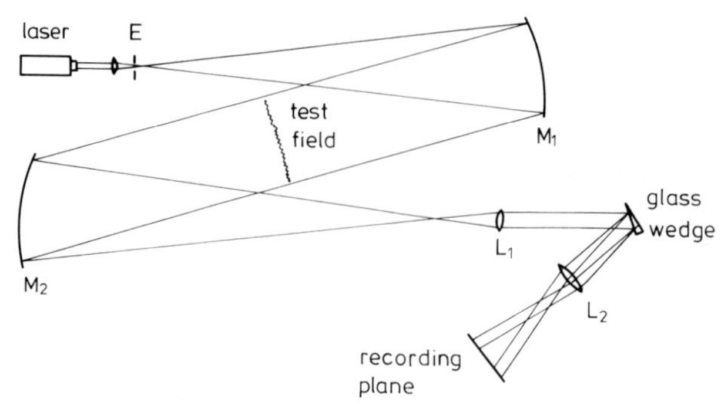

Fig. 3.44 Shearing interferometer using a wedge as the shearing element. (From Sandhu and Weinberg, 1972. Copyright 1972 The Institute of Physics.)

diverging rays that are polarized perpendicular to each other. This may be explained with the aid of Fig. 3.45. The optical axes of the two prism parts are oriented perpendicular to each other (marked ⊙ and '); α is the prism angle. An incident ray is separated in the first prism into the ordinary and the extraordinary ray; both rays coincide, when the angle of incidence is 90°, however they propagate in the prism at different speeds. Therefore, different indices of refraction n_o and n_e apply to the ordinary and the extraordinary ray, respectively. For quartz we have $n_e > n_o$ in the visible range of wavelengths. Because of the orientation of the optical axes the o ray of the first prism becomes the e ray in the second prism part, and vice versa. In the second half of the prism the two rays also exhibit a spatial separation; they leave the prism separated by an angle ε, which is found

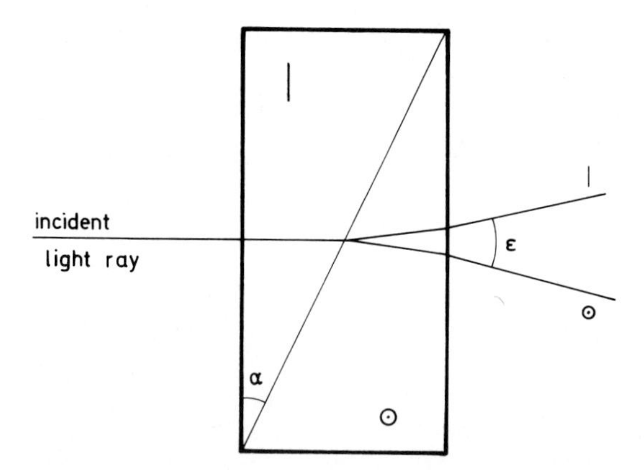

Fig. 3.45 Separation of light ray in a Wollaston prism.

TABLE 3.3

Refractive Indices for the Ordinary and Extraordinary Ray, n_o and n_e, in Calcite and Quartz at a Wavelength of 0.589 μm

	n_o	n_e
Calcite	1.6584	1.4864
Quartz	1.5442	1.5553

according to Snellius' law of refraction and by assuming a small prism angle α to be

$$\varepsilon = 2\alpha(n_e - n_o). \tag{3.42}$$

In a first-order approach this applies to normal as well as to oblique incidence of the primary ray. When leaving the prism the two polarized rays are deflected from the original light direction by an angle of $\pm\frac{1}{2}\varepsilon$, respectively. Most of these schlieren interferometers are operated with a Wollaston prism of relatively small prism angle α and, consequently, of small separation angle ε (see Table 3.3).

As for most shearing interferometers, the optical setup is very similar to a schlieren system. The knife edge is replaced here by the prism unit consisting of the Wollaston prism and two crossed polarizers (Fig. 3.46). The center of the Wollaston prism coincides with the focal point of the lens or spherical mirrors M (the "schlieren head") behind the test section (Fig. 3.47). The incident ray 1 is separated by the prism into two polarized rays, $1'$ and $1^{\odot}$, which form an angle ε with each other. (The symbols $'$ and $\odot$ indicate the polarization direction in the plane and perpendicular to the plane of the figure, respectively). A ray 2 including an angle ε with 1 in the converging light beam is separated into $2'$ and $2^{\odot}$. The separated ray $2'$ coincides with $1^{\odot}$, and both rays could interfere, if they had equal

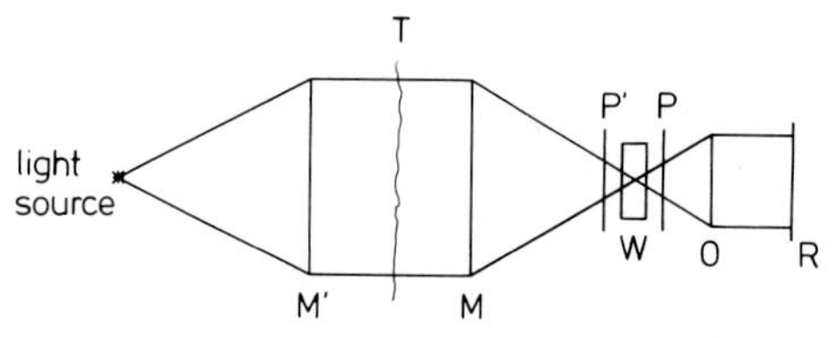

Fig. 3.46 Setup of Wollaston prism interferometer with parallel light through the test field. M and M′ are the lenses or spherical mirrors (M, schlieren head); P and P′, polarizers; W, Wollaston prism; O, camera objective; and R, recording plane (film).

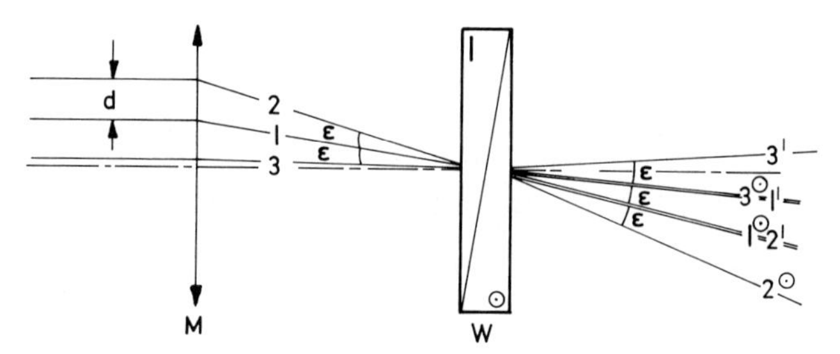

Fig. 3.47 Wollaston prism as the shearing element (interferometer unit).

direction of polarization. This additional condition is fulfilled by the polarizer P, which is rotated at 45° with respect to the axes of polarization of the Wollaston prism, while the polarizer P' in front of the prism provides equal intensity to the two interfering components. The same procedure applies to the rays 1 and 3 in Fig. 3.47 (i.e., 1′ and $3^{\odot}$ can interfere, etc.). The distance d by which conjugate rays are sheared or separated in the test field is $d = \varepsilon f_2$, with f_2 being the focal length of the mirror M ("schlieren head").

If the system is operated with a white light source, a second Wollaston prism may be used on the side of the light source in order to improve the optical coherence. In earlier descriptions of this interferometer, the role of this prism had been misinterpreted as to providing to the light beam the desired shear. That prism is not necessary when a coherent light source (laser) is used. Since the light emanating from a laser may exhibit a linear polarization, one might have to adjust the direction of this polarization with the directions given by the optical axes of the Wollaston prism.

If the center of the prism coincides with the focal point of the "schlieren head" M, and if there is a homogeneous density distribution in the test section, the two corresponding rays will not exhibit any phase difference. The field of view is free of interference fringes and covered with only one color ("infinite fringe width" alignment). A shift of the prism from this central position results in the formation of fringes of finite width. Now two corresponding rays propagate along paths of different optical lengths in the two prism parts; they leave the prism with different optical phase. The Wollaston prism can be displaced in horizontal and/or vertical direction; the corresponding phase shifts will be determined.

We assume first a horizontal shift of the prism of the amount w (Fig. 3.48). The width of the prism is 1. Neglecting the separation angle ε and assuming a small angle of incidence, we determine the path of a ray, which passes through the first prism half as ordinary and through the second half as extraordinary, to have the optical length of

$$l_1 = n_o(\tfrac{1}{2} - \alpha w y/f_2) + n_e(\tfrac{1}{2} + \alpha w y/f_2).$$

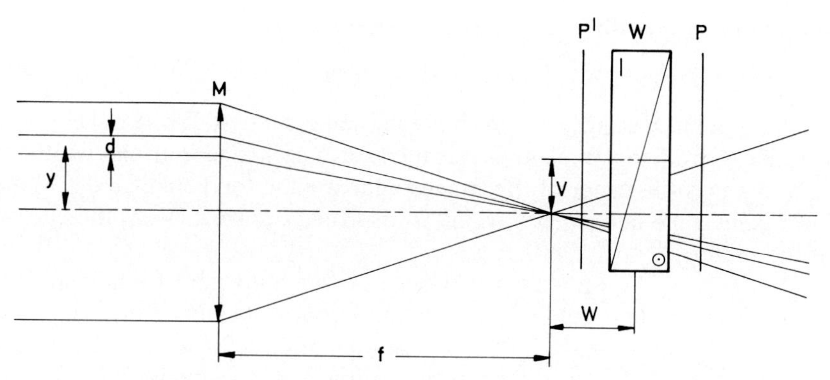

Fig. 3.48 Shift of the Wollaston prism in axial and normal direction.

y is the vertical position of this ray in the plane of the test section. In the same way we find this value for the conjugate ray, and the difference in optical path length of both rays is

$$\Delta l = 2\alpha(n_e - n_o)wy/f_2 = \varepsilon(w/f_2)y. \tag{3.43}$$

The corresponding optical phase difference is given by

$$(\Delta\varphi/2\pi)_w = (\varepsilon/\lambda)\,(w/f_2)y, \tag{3.44}$$

where λ is the light wavelength. According to Eq. (3.44) the phase difference of two interfering rays is proportional to the coordinate y of these rays in the test section. Their exact position in the test section is $y \pm \frac{1}{2}d$, respectively. The order of the interference fringes grows linearly with y, and we obtain a series of equidistant, parallel fringes, which, in our case, are oriented perpendicular to the plane of Fig. 3.48. In order to determine the fringe width S we consider two rays separated in the test section by a vertical distance of Δy. With a phase difference of $\Delta\varphi = 2\pi$ between these rays, Δy is equal to the fringe width S, for which we obtain, with Eq. (3.44),

$$S = \lambda f_2/\varepsilon w. \tag{3.45}$$

The fringe width decreases with the increasing distance of the prism from the focal point of M, and also with increasing separation angle of the Wollaston prism.

We now consider the case where the Wollaston prism is shifted in vertical direction by an amount v (Fig. 3.48). The two conjugate rays again traverse along optical ways of different lengths in the two prism halves; they leave the prism with a difference in optical path length of

$$\Delta l = 2\alpha(n_e - n_o)v = \varepsilon v, \tag{3.46}$$

and the corresponding optical phase difference is

$$(\Delta\varphi/2\pi)_v = \varepsilon v/\lambda. \tag{3.47}$$

All pairs of interfering rays now have the same phase difference. Hence, a vertical shift of the Wollaston prism causes no change in fringe width; it only changes the order of the fringes appearing in the field of view. With increasing v the achromatic fringe is displaced outwards from the center of the picture.

The form of the interference fringes or the equivalent fringe shift ΔS generated by a flow field of variable fluid density has been described by Eqs. (3.35), (3.36), and (3.37b). These equations refer to the case when the lateral shear d is produced in the y direction, and, consequently, the method is then sensitive to changes of $\partial n/\partial y$. By rotating the prism unit around the optical axis, one can measure the derivative of n in any wanted direction. Such a rotation also rotates the parallel fringe pattern of the finite fringe width mode. The derivative of n is always measured in a direction normal to this undisturbed fringe pattern (Fig. 3.49).

A characteristic feature of shearing interferometers is the formation of a double image of the edges of solid bodies. The formation of this double image is due to the blocking of one of the two conjugate rays by the solid wall. The double image is projected into the direction of the shear, and its width is $d \cos \alpha$, where α is the angle between the shearing direction and the normal to the wall. The exact position of the wall edge is in the middle of the gray zone or double image. Closely related with this phenomenon is the interference pattern generated by a shock wave. Since the density gradient through the shock is infinite, the linearized Eq. (3.35) cannot predict the fringe pattern in this case. Only those pairs of conjugate rays, which have one ray passing behind, and the other passing in front of the shock surface can contribute to the formation of the respective pattern (Fig. 3.50). One derives that the relative fringe shift (defined only in the lower part of Fig. 3.50) is

$$\left(\frac{\Delta S}{S}\right)_{\text{shock}} = \frac{K}{\lambda}\int_{\zeta_1}^{\zeta_2} \Delta\rho \, dz, \tag{3.48}$$

with $\Delta\rho$ being the density jump through the shock, here assumed to be a surface y = const. Equation (3.48) has the same form as a respective equation for a reference interferometer, however, the form of the pattern is quite different in the two cases.

Since the heat transfer coefficient is directly proportional to the gradient of temperature, shearing interferometers are of particular interest for convective heat transfer studies. For, the pressure can be taken as constant in most of these cases, and the fringe pattern then reflects the distri-

Fig. 3.49 Candle flame visualized with a Wollaston prism interferometer at finite fringe width. (From W. Erdmann and W. Merzkirch, Ruhr-Universität Bochum.)

Fig. 3.50 Representation of a plane shock wave with a shearing interferometer at finite fringe width. Direction of fringes is (a) parallel and (b) oblique to the plane of the shock.

bution of the temperature gradient (Fig. 3.51). The principles of the Wollaston prism interferometer have been revised several times under the aspect of measuring heat transfer coefficients (Sernas and Fletcher, 1970; Black and Carr, 1971; Small *et al.*, 1972). A great number of applications have been reported for measuring convective flows (Aung *et al.*, 1972; Black and Norris, 1974; Miller and Gebhart, 1978; Bühler and Oertel, 1982; Oertel, 1982), mass transfer (Stirnberg *et al.*, 1983; Carlomagno, 1985), and plasma flows (Kogelschatz, 1974). Flack (1978b) investigated the errors introduced by the misalignment of the surfaces of test objects. Branston and Mentel (1976) present a theory of the Wollaston prism interferometer, which is not based on geometrical optics. The case of weak phase objects, when the overall fringe variation is less than one fringe width, has been studied by Smeets (1970). It turns out that, in the infinite fringe width mode, the variation of luminance in the field of view is described by a formula identical with the respective equation for a schlieren system, so that the close relationship between the two methods becomes again evident.

3.3.4. Light Sources, Recording Methods, Evaluation

The quality of the recorded interferometric informations is affected, besides other influences, by the characteristics of the components of the optical system: light source, mirrors, lenses, windows, mechanical supports, recording system. We will only briefly discuss the role of the light sources and the recording systems, and the reader is referred to the perti-

Fig. 3.51 Wollaston prism interferograms of the natural convective flow along a heated, vertical plate. Temperature difference ΔT_w between wall and ambient air is (a) 20°C and (b) 63°C, respectively. (From Shu Ji-zu, Ruhr–Universität Bochum and Academia Sinica, Beijing, and W. Merzkirch.) ▶

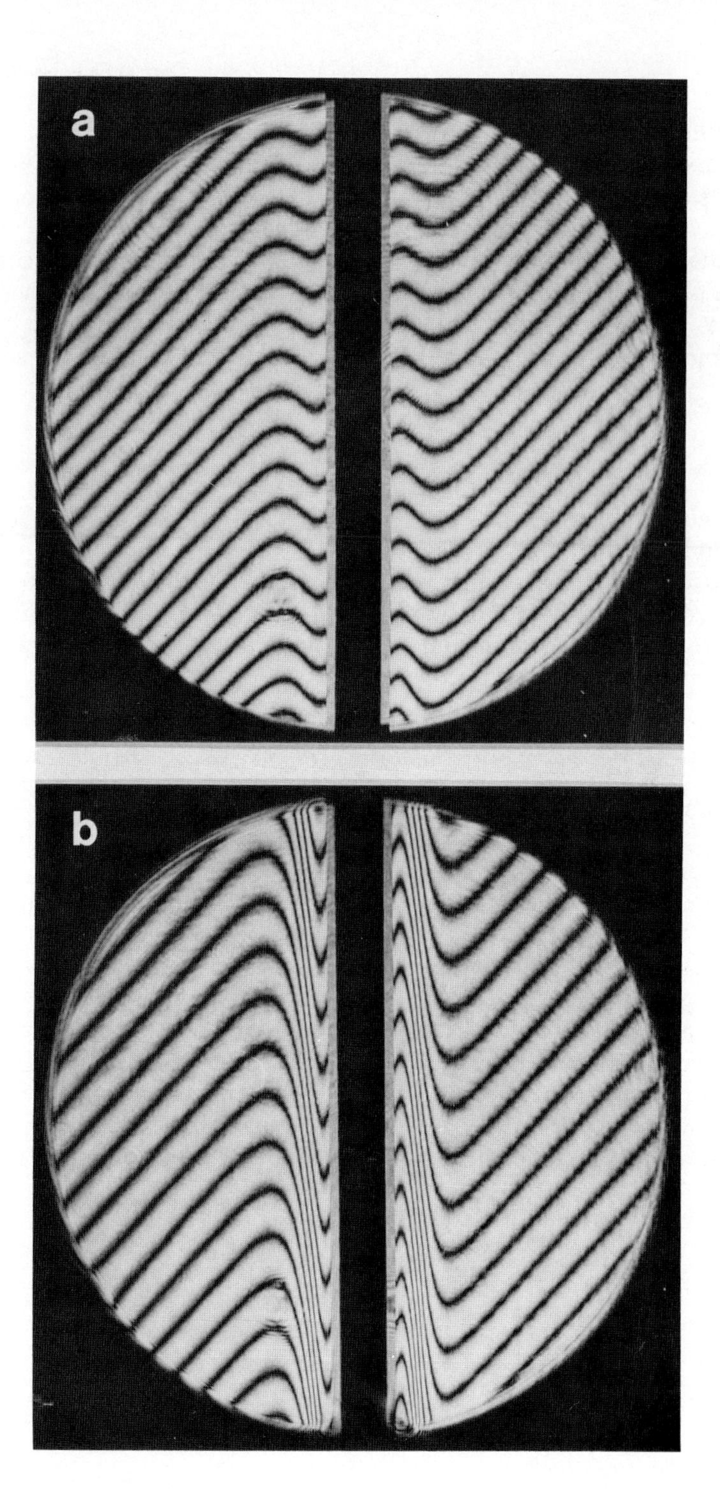
a
b

nent literature for more details (e.g., Hugenschmidt and Vollrath, 1981). We can distinguish between thermal (white) and laser (monochromatic) sources. This distinction is of particular relevance for interferometric testing, because the laser light has a much higher degree of optical coherence than the light from thermal sources. Optical coherence, necessary for the interference of light waves, can be measured in terms of coherence time, τ_{coh}. This quantity, which is desired to be large, is related to the frequency range of the emitted light, $\Delta\nu$, by the order-of-magnitude relation

$$\tau_{coh} \sim 1/\Delta\nu .$$

The high degree of (time-dependent) coherence of laser light is thus expressed by the very small value of $\Delta\nu$.

A further question is under which conditions the light emitted from an extended source of width d and of aperture angle ϑ (Fig. 3.52) can still interfere. The answer is found from Young's classical experiment, and the requirement for this spatial coherence is

$$d \sin \vartheta \ll \lambda/2 .$$

The high degree of spatial coherence of laser light becomes evident by the extreme parallelism of the laser beams, which have an aperture angle of the order of 1×10^{-3} rad.

Another possible classification of light sources is the time of light emittance. One can distinguish between continuous sources (e.g., mercury lamps and continuous lasers), and sources producing light pulses, either a single pulse or regular series of pulses (strobe light). The width of the pulses ranges from milliseconds (flash lamps), to microseconds (electric spark sources) and nanoseconds (pulsed lasers). High-speed pulse sources have been extensively discussed in relation to high-speed photography (see, e.g., Vollrath and Thomer, 1967, and the proceedings of the various congresses on high-speed photography). Since the application of a

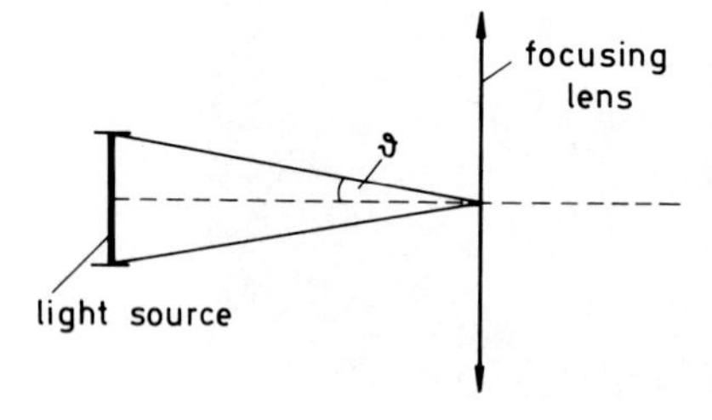

Fig. 3.52 Definition of the aperture angle ϑ of an extended light source.

pulse laser is also a matter of costs, the development of short, inexpensive thermal sources with satisfactory coherence characteristics has always been of great interest (see, e.g., Preonas and Swift, 1970; Parks and Sumner, 1978; Miyashiro and Grönig, 1985).

In order to expand the thin laser beam for being used in an optical visualization system, one applies a beam expanding optics consisting of an objective of very short focal length (~ 1 cm) and a pinhole ("spatial filter") placed in the focal plane of the objective (Fig. 3.53). The purpose of the pinhole is to clean the laser light from diffractive noise. Its diameter is approximately equal to the diameter of the zeroth diffractive order of the light source image in that focal plane. It is known that all the information on possible images, here: the "speckles" representing the noise, is contained in the higher diffractive orders, so that the beam after having passed the spatial filter, has homogeneous brightness. In the early days of laser technology, the application of laser sources to flow visualization systems had not only resulted in definite advantages, but also generated problems that were unknown until then, namely the diffractive noise, which arises from the high degree of coherence of laser light. The role of laser sources in interferometry has been reviewed, therefore, on several occasions (e.g., Tanner, 1966b; Oppenheim *et al.*, 1966).

One of the definite advantages of laser light in interferometry is the great number of distinct, clear interference fringes. Kinder (1946) derives for the total number of fringes, N, obtainable on either side of the achromatic (i.e., zeroth order) fringe $N \simeq 1/\vartheta^2$ with ϑ being the aperture angle of the light source according to Fig. 3.52. A minor shortcoming of this identity of nearly all fringe orders is the difficulty to identify the same fringe on either side of a shock wave and, therefore, to determine the exact fringe shift in this case. Grönig (1967) proposes to illuminate a small slit of the field of view with a white source, so allowing for an easier identification of the density jump.

The wavelengths of the ruby and the helium–neon laser are in a spectral range where the sensitivity of normal photographic material is rather low (Fig. 3.54). The argon ion laser has more favorable characteristics.

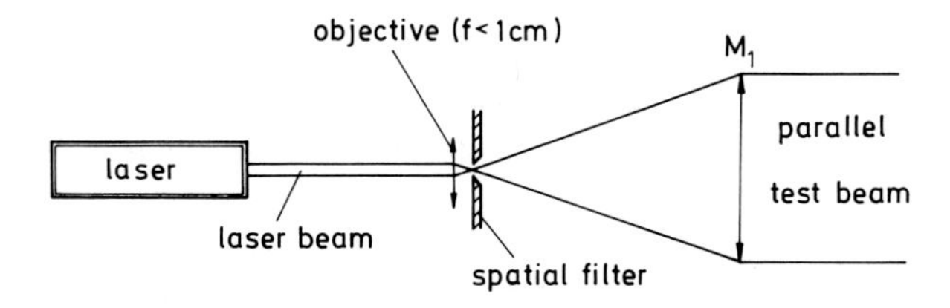

Fig. 3.53 Beam expander and spatial filter applied to a laser beam.

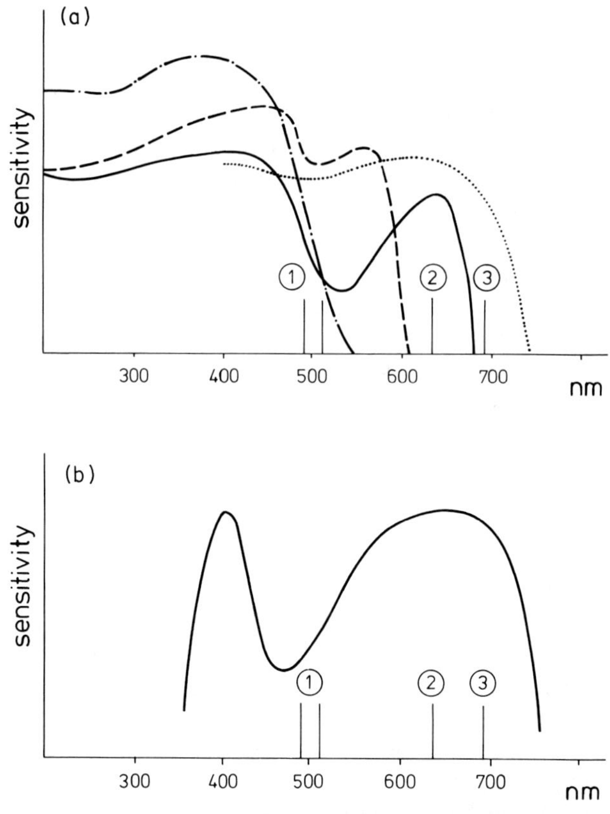

Fig. 3.54 Spectral sensitivity of (a) four typical photographic emulsions and of (b) an emulsion used for holographic recording. Sensitivity is given in arbitrary units. Also indicated are wavelengths of radiation from an ① argon ion laser, ② He/Ne laser, and ③ ruby laser. It is seen that some emulsions are totally insensitive for the red light from a ruby laser. (Data taken from brochures of Kodak and Agfa–Gevaert.)

Besides photographic and holographic materials, television cameras and arrays of photodiodes can be used for the recording; for details see Hugenschmidt and Vollrath (1981). The latter two methods are of interest if the pattern of interference fringes should be digitized for automated evaluation. In general, the evaluation consists of providing the interferometric information, either in form of the optical path length difference $\Delta l/\lambda$ [Eq. (3.33)] or in form of the fringe shift $\Delta S/S$ [Eq. (3.37)] in the recording plane (x, y), and to determine from these quantities the respective density distribution. The information is available along the interference fringes, and in between the fringes it must be determined by interpolation. The first step to extract the information is to locate the fringe centers. This can be done manually, or it can be digitized by scanning the fringes with a densitometer whose reading is given to a computer (Mastin and Ghiglia, 1985). The digitized fringe contours can be approximated by geometrical curves (Kim, 1982), and appropriate interpolation allows for a continuous representation of the information in the coordinates x, y. Methods for automated fringe reading and data reduction have been developed in different fields of application of interferometry (e.g., Boxman and Sloan, 1978; Cline *et al.*, 1982; Robinson, 1983; Boyd *et al.*, 1985). Particular difficulties arise from shock waves due to the abrupt changes in the fringe contour (Ben-Dor *et al.*, 1979). The automated methods by which the evaluation becomes independent of subjective measurements certainly improve the repeatability but not necessarily the reliability of the results.

3.3.5. *Three-Dimensional Density Fields*

3.3.5.1. *Axisymmetric Fields*

In this section we discuss how the two-dimensional information that is obtained from an interferogram can be used for determining the density in a three-dimensional flow field. We begin with a special case and assume that the flow field has rotational symmetry. This case occurs quite frequently, for example, in form of axisymmetric jets, flames, free convection along a vertical heated cylinder, or supersonic flow over axisymmetric configurations without angle of attack. However, for applying the analysis to be discussed here, it is required that the flow in those situations be laminar.

We take the x axis as the axis of rotational symmetry, and the z axis as the optical axis, that is, the light propagates in z direction through the axisymmetric test field. (Fig. 3.55). The refractive index and the fluid in the flow depend on the coordinates x and $r = \sqrt{y^2 + z^2}$. These two coordinates must be introduced into the respective equations describing the information, which is available from an interferogram. Since we want to discuss reference beam and shearing interferometry, as well as finite and infinite fringe width modes simultaneously, a data function D and a refractive index function R are introduced by

$$D(x, y) = \mathbf{c} \int_{\zeta_1}^{\zeta_2} R(x, y, z)\, dz \tag{3.49}$$

The forms of D, R, and of the constant $\mathbf{c}$ are found by comparison of Eq. (3.49) with either one of the respective equations [Eqs. (3.32), (3.34),

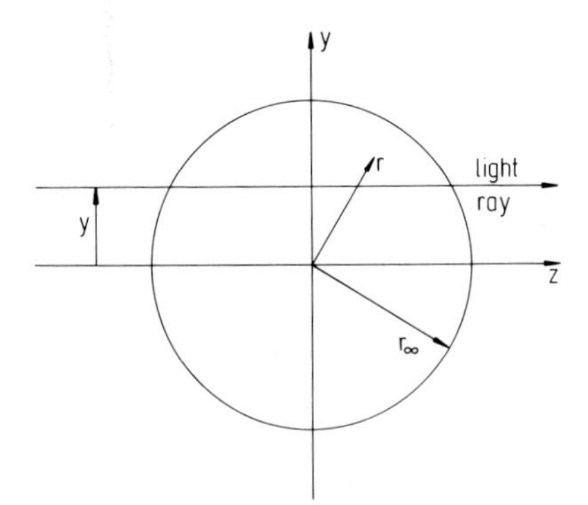

Fig. 3.55 Light ray passing through an axisymmetric test field at position y = const. The x axis, normal to the plane of the figure, is the axis of rotational symmetry.

(3.35), and (3.37a,b). The data function D designates either the relative fringe shift $\Delta S/S$ or the fringe order (optical phase difference Δl divided by the wavelength λ) in a point (x, y) of the interferogram. With the line element,

$$dz = r\,dr/\sqrt{r^2 - y^2},$$

Eq. (3.49) is transformed into

$$D(x, y) = \int_{r=y}^{r_\infty} R(x, r)\,\frac{d(r^2)}{\sqrt{r^2 - y^2}}. \tag{3.50}$$

It is assumed that the axisymmetric test field is bounded by the radius r_∞, and that n, ρ and R are constant for $r \geqq r_\infty$ (i.e., outside of the test field). The forms of the functions D and R, into which all constants are incorporated, are listed in Table 3.4. Equation (3.50) and Fig. 3.55 show that all refractive index values with $r > y$ contribute to the variation in optical phase of the ray propagating at the position y. An apparent difficulty for the further analysis arises from the singularity at $r = y$.

A direct approximation for evaluating the integral in Eq. (3.50) has been described by Schardin (1942). A circular cross section of the test field of radius r_∞ is subdivided into N annular zones of width h (Fig. 3.56); r_μ designates the outer radius of the μth zone, and one has

$$O = r_0 < r_1 < r_2 < \cdots < r_{N-1} < r_N = r_\infty,$$

and $|r_\mu - r_{\mu-1}| = h$. It is assumed that the refractive index function R is constant within each annular zone, and that it changes discontinuously from one zone to the next. This means that the density is considered as constant within each zone if one evaluates a reference beam interferogram, and that the density gradient is taken as constant in each zone in the case of a shearing interferogram (Merzkirch and Erdmann, 1973). The

TABLE 3.4

Data Function D and Refractive Index Function R in the Evaluation of Axisymmetric Refractive Index Fields (Eq. 3.50)

	Reference beam interferometer	Shearing interferometer
Infinite fringe width alignment	$D = \dfrac{\Delta l(x, y)}{\lambda}$	$D = \dfrac{1}{yd}\dfrac{\Delta l(x, y)}{\lambda}$
	$R = \dfrac{1}{\lambda}\{n(x, r) - n_\infty\}$	$R = \dfrac{1}{\lambda}\dfrac{\partial n(x, r)}{\partial(r^2)}$
Finite fringe width alignment	$D = \dfrac{\Delta S(x, y)}{S}$	$D = \dfrac{1}{yd}\dfrac{\Delta S(x, y)}{S}$
	$R = \dfrac{1}{\lambda}\{n(x, r) - n_\infty\}$	$R = \dfrac{1}{\lambda}\dfrac{\partial n(x, r)}{\partial(r^2)}$

principle of the procedure that solves the integral equation in form of a step-function is indicated by the two rays in Fig. 3.56. Ray 1 passes only through one annular zone, and the optical signal it delivers is sufficient for determining the value of R_N, provided of course, that the entrance and exit coordinates of the ray in the zone are known. Since R_N has been determined, ray 2 also passes through only one zone of unknown value of R. An equation can be written relating the optical signal of ray 2 to the known contribution of R_N and the contribution of R_{N-1} that is to be determined. This procedure continues until one arrives at the ray through the center of the circular cross section.

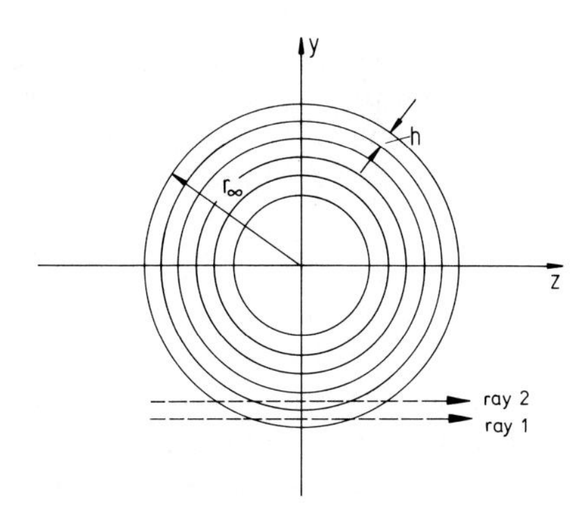

Fig. 3.56 Subdivision of the axisymmetric test field into annular zones of width h.

For the mathematical formulation of Schardin's approximative solution it is assumed that discrete values of the data function, $D_\mu = D(y_\mu - 1)$, are known at the inner edge of each annular zone. For determining discrete values of the refractive index function in the middle of each zone, $R_\mu = R(r_{\mu-1} + h/2)$, the integral [Eq. (3.50)] is approximated by the following summation

$$D_\mu = D(y_{\mu-1}) = \sum_{i=\mu}^{N} R\left(r_{i-1} + \frac{h}{2}\right) \int_{y_{i-1}}^{y_i} \frac{d(r^2)}{\sqrt{r^2 - y_{\mu-1}^2}}. \tag{3.51}$$

Algorithms for the further handling of this equation are available, but unfortunately in form of internal laboratory reports the access to which is usually difficult (e.g., Gorenflo, 1964; Oudin and Jeanmaire, 1970). Equation (3.51) shows that all discrete values R_i with $\mu < i < N$ contribute to the generation of the data value Δ_μ. In the notation of Gorenflo (1964), Eq. (3.51) is replaced by

$$D_\mu = \sum_{i=\mu}^{N} \alpha(\mu, i) R_i,$$

where the coefficients α are determined by solving the integral in Eq. (3.51):

$$\alpha(\mu, i) = 2h\{\sqrt{i^2 - (\mu - 1)^2} - \sqrt{(i - 1)^2 - (\mu - 1)^2}\}.$$

The recursive solution of the system of equations for R is

$$R_\mu = \frac{1}{2h\sqrt{2\mu - 1}} \left\{D_\mu - \sum_{i=\mu+1}^{N} \alpha(\mu, i) R_i\right\}. \tag{3.52}$$

This step-by-step solution starts at the outer edge of the axisymmetric field with

$$R_N = R\left(r_{N-1} + \frac{h}{2}\right) = \frac{D_N}{2h\sqrt{2N - 1}},$$

$$R_{N-1} = \frac{h}{2}\sqrt{2N - 3}\,\{D_{N-1} - R_N\{2\sqrt{N - 1} - \sqrt{2N - 3}\}\}.$$

$$\vdots$$

Values $R(r)$ with $r_\mu < r < r_{\mu-1}$ can be derived by linear interpolation. Generally, the accuracy of the method increases with the value of N.

Another approach for resolving the axisymmetric refractive index distribution makes use of the fact that the integral [Eq. (3.50)] is of the classical Abel type. The so-called Abel inversion, derived from the theory of Laplace functions, furnishes for the solution of this integral equation

$$R(r) = -\frac{1}{\pi}\frac{d}{d(r^2)}\int_r^{r\infty}\frac{D(x, y)}{\sqrt{y^2 - r^2}}\,d(y^2). \tag{3.53}$$

The integration has to be performed over the known function $D(x, y)$. Since the data, at given x = const, is only available in form of a set of discrete data points $D(x, y_\mu)$, the function $D(x, y)$ should be approximated by a suitable analytic expression, or by a linear combination of suitable functions that match $D(x, y_\mu)$ at the discrete experimental points y_μ (Solignac, 1965; South, 1970). The advantage of automated data evaluation as described in Section 3.4 becomes visible. Again, details for computer programs and approximative algorithms are described in a number of laboratory reports (e.g., Kean, 1961; Oudin and Jeanmaire, 1970).

Use has been made of the Abel inversion also in other fields of optical experimentation (e.g., spectroscopy, radio-astronomy, and plasma diagnostics). The procedure has the disadvantage that the necessary differentiation $d/d(r^2)$ of the experimental data in Eq. (3.53) is usually associated with the generation of considerable errors. As shown by Kean (1964) and exemplified (e.g., by Stricker (1984)), this disadvantage is avoided for shearing interferometry or other methods depending on the measurement of the refractive index gradient. If Eq. (3.53) is written for this particular case, the two differentations $d/d(r^2)$ appearing on both sides of the equation are eliminated by a simple integration, thus yielding

$$n(x, r) - n(x, r_\infty) = -\frac{\lambda}{\pi d}\int_{r=y}^{r\infty}\frac{1}{y}\frac{\Delta S(x, y)}{S}\frac{d(y^2)}{\sqrt{y^2 - r^2}}. \tag{3.54}$$

Equation (3.54) is given for a shearing interferometer in the finite fringe width alignment. The equation presents the surprising result that, for an axisymmetric test field, shearing interferometers deliver the absolute value of the refractive index or density.

Particular difficulties arise if the test field contains conical or cylindrical discontinuity surfaces, e.g., shock waves. According to Weyl (1954) one may split the refractive index function as well as the data function into a "singular" portion, which accounts for the discontinuous density jump across such a surface, and into a "reduced" portion describing the continuous density change. The "singular" portion of the data function D is determined from the respective fringe shift at the discontinuity surface, while the "reduced" portion of D is found by subtracting the "singular" portion from the measured overall data function. The "reduced" portion of the refractive index function can then be determined with one of the procedures described earlier. The effect of such discontinuities on the reconstruction of the refractive index field, both for axisymmetric and nonsymmetric cases, has been discussed by Maruyama *et al.* (1977). The analysis of axisymmetric supersonic flow fields with conical shock waves has been reported, e.g., by Bennett *et al.* (1952) and Solignac (1968).

3.3.5.2. Tomography

If the refractive index field has no symmetry, the refractive index distribution $n(x, y, z)$ cannot be determined from just one interferogram providing the data $D(x, y)$. This completely three-dimensional situation is illustrated in Fig. 3.57, which gives a cross section at x = const through the test field. We assume that the field can be subdivided into N segments each of which is assigned to a constant value of the refractive index function R, so that $R = R_i$ in the ith segment. The problem is to determine N unknown quantities $R_1 \ldots R_N$, and this requires defining and solving a set of N independent equations. A light ray traversing the test field carries information on the R_i-values of m segments, with $m < N$. This information is expressed by a value D_j of the data function $D(x, y)$. Equation (3.49) written for this one light ray is then just one equation of the aforementioned set of N equations. For this simplified discussion we might

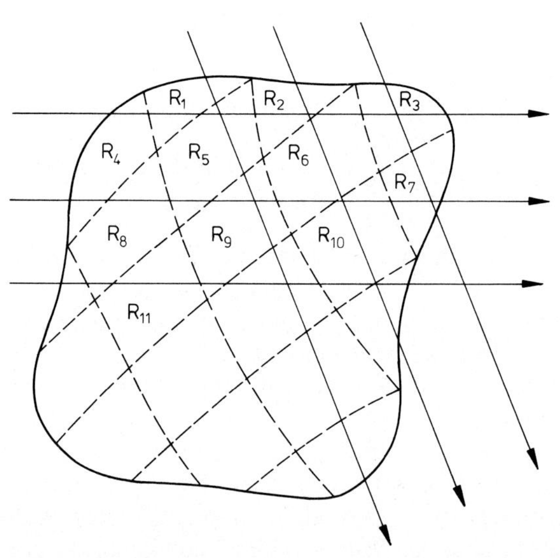

Fig. 3.57 Subdivision of the cross section x (= const) of a three-dimensional test field into segments with constant values R_i of the refractive index function R. The information obtained from the six light rays shown in the figure delivers six equations for the evaluation procedure.

assume that the integral in Eq. (3.49) can be approximated by a summation, so that the first equation of the system (which is linear now) reads

$$D_1 = \sum_{i=1}^{m_1} R_i s_i$$

where m_1 is the number of segments through which light ray 1 passes in the test field, and s_i is the geometrical length covered by the light ray in the ith segment. Additional equations arise in expressing the paths of other light rays traversing the field at different positions and parallel to ray 1.

From Fig. 3.57 we may conclude that one beam of parallel light rays cannot yield sufficient information to complete the system of N equations. It is therefore necessary to utilize the information from other rays, which pass through the test field in a different direction, that is, to measure the test field with the interferometer under different viewing directions. In the case of an arbitrary three-dimensional refractive index field, it is necessary to observe the object over an angular range of 180°. This value and the number of viewing directions required decrease if the test field has some kind of symmetry. This means, in terms of our simplified model, that some segments have the same R_i values, so that the number of unknowns and the number of needed equations are reduced. As we have seen, only one viewing direction is sufficient in the case of rotational symmetry.

The direct analogy between our simplified model and Schardin's approach to the axisymmetric case is apparent. Such discretisized methods that result in systems of large numbers of linear equations have been applied by Belotserkovsky (1968) and Schwarz and Knauss (1982). As for axisymmetric flow, another approach utilizing inversion methods is possible. This is what is usually designated as *tomography*. The problem is to reconstruct a three-dimensional object field from a set of two-dimensional (plane) projections taken in various directions. This problem, which was formulated mathematically already in 1917 by Radon, arises in a number of different fields (e.g., x-ray diagnostics, electron microscopy, radio astronomy) and a great amount of special literature is therefore available. The state of the art has been reviewed by Herman (1980). The relationship between Radon and Abel inversions has been investigated by Vest (1974), and the particular application to the reconstruction of three-dimensional refractive index fields has been discussed by Sweeney and Vest (1973).

With the notation shown in Fig. 3.58 the equation to be inverted is

$$D(x, \eta, \alpha_i) = \int_{\zeta_{i1}}^{\zeta_{i2}} R(x, y, z)\, d\zeta \tag{3.55}$$

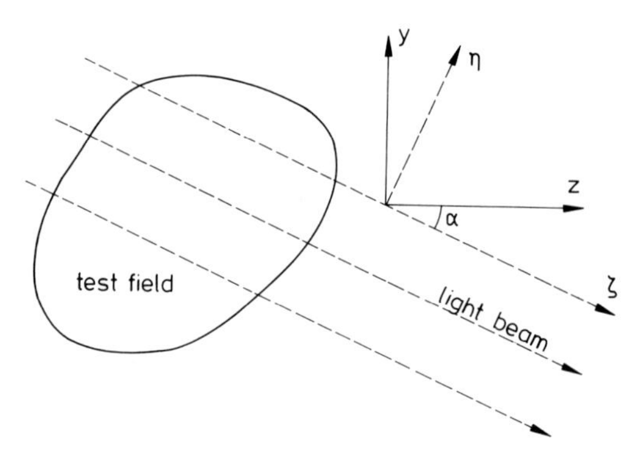

Fig. 3.58 Light trajectories and coordinate system in a cross section $x(= \text{const})$ of a three-dimensional refractive index field.

where ζ_{i1} and ζ_{i2} are the entrance and exit coordinates of the light rays in the ith projection designated by the angle α_i. $D(x, \eta, \alpha_i)$ is the interferometric data that is collected, for this particular projection, in the x, η-plane (normal to the ζ direction). Matulka and Collins (1971) represent the refractive index function by a series of generating function $U_{m,k}$:

$$R(x, y, z) = \sum_{m=1}^{\infty} \sum_{k=1}^{\infty} a_{m,k} U_{m,k}(x, y, z),$$

where the series coefficients $a_{m,k}$ are determined by orthogonalization. Data must be recorded over a range of 180° for the viewing angle. The inversion is then of the form

$$R(x, y, z) = \sum_{m=1}^{\infty} \sum_{k=1}^{\infty} \left\{ \int_{-\pi/2}^{\pi/2} \int_{-\infty}^{\infty} D(x, \eta, \alpha) H_{m,k}(\eta)\, d\eta\, d\alpha \right\} U_{m,k}(x, y, z),$$

where $H_{m,k}$ are Hermite polynomials. This procedure was first tested with a slightly asymmetric free jet, and it was then applied to the study of supersonic flow over circular cones at angle of attack (Jagota and Collins, 1972), to transonic corner flow (Kosakoski and Collins, 1974), and to the mixing in a buoyant plume (Witte and Mantrom, 1975).

A method employing the inversion characteristics of Fourier transforms has been proposed by Rowley (1969) and was brought to a form more feasible for numerical analysis by Junginger and van Haeringen (1972). Zien *et al.* (1974, 1975) applied this evaluation method to a large-scale wind tunnel experiment, the supersonic flow over a cone at yaw. Both Zien *et al.* (1974) and Vest and Prikryl (1984) discuss the problem of reconstructing the test field from data having incomplete projections because an opaque object, here: the model in the wind tunnel, is blocking off a part of the test beam. The development of optical tomography progresses with the advancement of computer science, and appropriate software is now available for reconstructing complex flow fields (see, e.g., Snyder and Hesselink, 1984). Finally, it should be mentioned that tomography is also playing a role for flow diagnostics based on absorption measurements (e.g., Santoro *et al.*, 1981).

3.3.6. Strong Refraction Effects

A refractive index gradient normal to the direction of the incident light causes the light rays to be bent out of their original direction of propagation. If this bending due to refraction is strong, Eq. (3.32) and the subsequent analysis of interferometry becomes invalid, because these equations had been derived under the assumption of straight light propagation through the test field. Such strong refraction effects may occur in combustion and convective heat transfer with strong temperature gradients in the fluid, and they can be of particular influence when the interferometric experiments are performed in liquids. For, the specific refractivity $(n - 1)$ is not a very small quantity for liquids. The refractive bending of light rays is an error source also in other fields of optical experimentation, and it has been studied, for example, in laser–Doppler anemometry for estimating the error in the definition of the measuring point (Schmidt *et al.*, 1984). The effect is visualized in Fig. 3.59 with a laser beam entering horizontally a glass tank, which is filled with stratified saltwater (i.e., brine with a strong vertical gradient of the salinity). The beam, which is visible due to light scattering is bent toward the direction of increasing value of the refractive index.

Full inclusion of the refraction effects requires developing, for the interferometric methods, a new theory starting with the complete Eq. (3.14), and not neglecting thereby the first derivatives (dx/dz), (dy/dz). Evaluation of an interferogram then becomes a problem in which not only the refractive index is unknown, but also the geometrical paths of the light rays that have contributed to the formation of the interference pattern. The problem has been looked after for a number of special cases, and most of the approaches are numerical studies that demonstrate, by means of optical ray tracing, the difference between strong refraction and the refraction-free situation.

The thermal boundary layer close to a heated, vertical flat plate, the concentration boundary layer close to a plane electrode in an electrolytic reaction, the stratified solution of saltwater are examples of two-dimen-

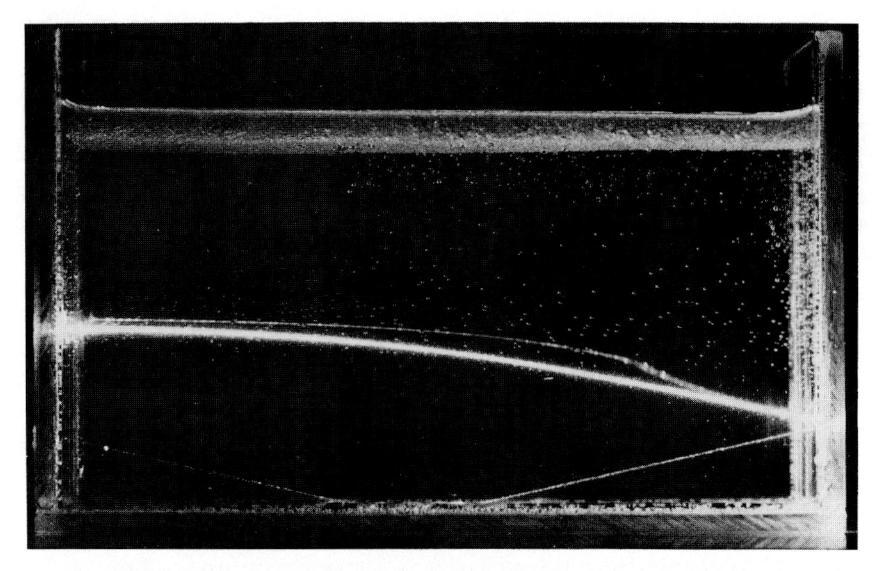

Fig. 3.59 A horizontal laser beam enters, from the left, a glass tank filled with stratified saltwater. Strong refraction causes the laser beam to be bent towards the higher densities in the fluid. The direct beam and reflections from the tank walls are seen due to sidewards scattering. (Courtesy of F. Peters, Universität Essen.)

sional test fields with $n = n(x, y)$, where the strong refraction is effective only in one direction (in the y direction). The refractive bending of light may be studied in a plane x = const. If one knows in this plane the exact distribution of the refractive index, $n = n(x, y)$, one may calculate the exact trajectories of the light rays from Eq. (3.14). If one knows only the algebraic form of the refractive index function, one may predict certain characteristics of the light trajectories. This is the scope of a number of approximate methods that only apply for small values of the light deflection. Assuming a linear distribution of the refractive index in the plane x = const, one derives that the light trajectory through the test field is a parabola (Grigull, 1963; Howes and Buchele, 1966; Brand and Grigull, 1974; Anderson *et al.*, 1975, see also Fig. 3.60). The refractive index field could then be reconstructed from the interferogram by making use of this "correction parabola" in the respective analysis. The correction parabola is in error when the refractive index gradient is not small or when the width of the test field (in z direction) is not short. Anderson *et al.* (1975) and Mowbray (1967) have shown that under the assumption of a linearly varying refractive index one may obtain a solution of the complete Eq. (3.14), and determine the exit coordinate, the slope, and the phase of the

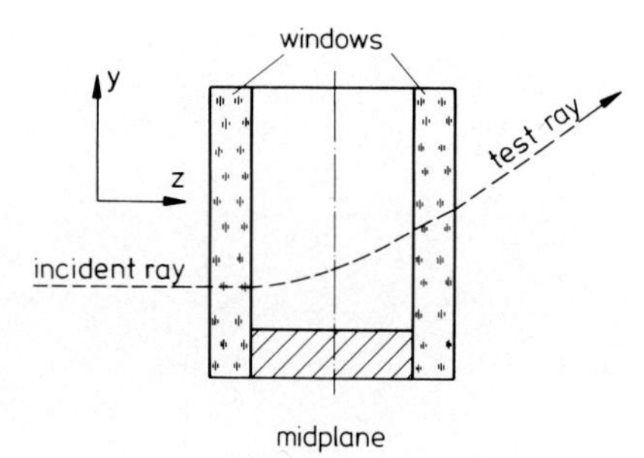

Fig. 3.60 Light trajectory in a test field with strong refractive index gradient $\partial n/\partial y$. The test section is bounded by plane windows.

rays leaving the test field. By comparing the exact with the approximate solution they find, for special cases like a thermal boundary layer, a criterion, which gives the greatest allowable refractive index gradient so that the correction parabola can still be used. MacLarnon *et al.* (1975) study the problem of strong refraction with the application to concentration boundary layers in liquids. In principle, their procedure consists of calculating an interferogram for an assumed algebraic model of the refractive index distribution, and comparing the calculated with the real interference pattern. By varying the values of a set of parameters in the model one tries to minimize the differences in the interferometric patterns.

In calculating the interference pattern for a given refractive index distribution in the test field, one must account for the imaging process in the interferometer, and for possible refractive distortions caused by the test section windows. The latter effect has been shown to be negligible in most cases (Kahl and Mylin, 1965; Howes and Buchele, 1966). Due to strong refraction one obtains a distorted image of the test field in the recording plane. This distortion is minimized by focusing onto the recording plane a plane in the test field situated between the center and a position at one-third the field width away from the exit plane into the test section (Mehta and Worek, 1984).

Attempts have been made for extending the investigation of strongly refracting test fields beyond the one-dimensional case to axisymmetric fields with a strong gradient of the refractive index in radial direction and negligible refraction in the direction of the axis of rotational symmetry. Examples are the thermal boundary layer along a vertical heated cylinder, the concentration boundary layer around a cylindrical electrode in an electrolytic reaction, or the plasma of an electric arc discharge of circular cross section. With n being an increasing function of the radial coordinate r, the light rays are bent outward in such a field (Fig. 3.61). The axisymmetric problem has been studied with numerical examples and computer simulations (Hunter and Schreiber, 1975; Vest, 1975; Montgomery and

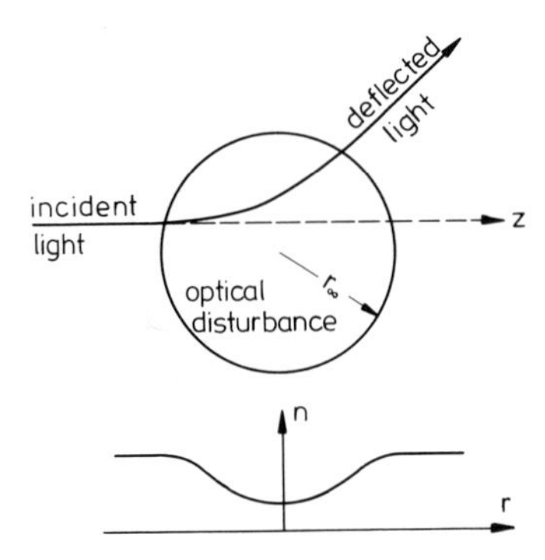

Fig. 3.61 Light trajectory in an axisymmetric test field with strong refractive index gradient in radial direction, $\partial n/\partial r$.

Reuss, 1982). Vest derives an analytic inversion, and he also proposes an approximate inversion scheme, which appears to be more feasible for practical handling. Montgomery and Reuss find that the refractive errors are minimized if the recording plane is imaged to the center plane of the flow field. An iterative algorithm for the tomographic reconstruction of asymmetric, strongly refractive fields has been described by Cha and Vest (1981).

3.3.7. Interferometry at High Temperatures

The application of optical interferometry to a gas mixture delivers the refractive index of the mixture, n, which is coupled with the partial densities of the components in the mixture, ρ_i, through Eq. (3.8). Since the local concentration values of the components are not known *a priori,* it is in general not possible to derive from the interferogram the density distribution or related quantities, like temperatures or concentration. Simple solutions to this problem, which is of particular interest in the study of flames, are available only for a few special cases. Weinberg (1963) has shown that the Gladstone–Dale constant of a gas mixture depends on the molecular weight of the mixture and on the polarizability of the atoms in the mixture, but not on their chemical bonding in molecules. For the premixed combustion of methane, where the number of moles does not change during the reaction, it follows that the Gladstone–Dale constant has the same value throughout the mixture so that the measurement of temperatures in a methane flame becomes possible (Schultz–Grunow and Wortberg, 1961).

In all other cases, when the aim is to measure temperatures in flames, certain information in addition to the interferogram is required. This can be either experimental data measured by additional means, or information from a respective theoretical model. Such a model, that reduces the number of unknowns to be determined, has been proposed by Pandya and Weinberg (1964) for a diffusion flame, and it has been applied by Stirnberg *et al.* (1983) to a laminar free jet with strong diffusion. This flow can be modelled with the principle of "conserved properties", so that all parameters describing the flow (density, temperature, refractive index, etc.) follow the same functional behavior, and it is necessary to measure only one quantity, here: the refractive index, from which the other parameters of interest can be derived.

Another possible approach to the interferometric study of the flow of a gas mixture with unknown concentration would be the recording of interferograms with different wavelengths. The number of used wavelengths would have to be equal to the number of components in the mixture, say N. This would yield a set of N equations of the form

$$n(\lambda_j) - 1 = \sum_{i=1}^{N} K_i(\lambda_j)\rho_i \tag{3.56}$$

where n is the refractive index of the mixture at one of the N used wavelengths λ_j, the K_i are the Gladstone–Dale constants, and ρ_i the partial densities of the N individual components. Unfortunately, the K_i values for most gaseous components are so weakly dispersive, that the interferograms taken at different wavelengths in the visible range differ not enough for providing independent information. There is only one field of application where the principle of multiple-wavelengths interferometry has been used successfully, namely plasma flows. In the discussion of Eq. (3.10) is has been shown that the Gladstone–Dale constant for the electron gas is strongly dispersive, so that, with the application of two-wavelengths interferometry to plasma flows, one may determine the two unknowns in Eq. (3.10), ρ and N_e. For this purpose, Hugenschmidt and Vollrath (1970) operate a ruby laser at $\lambda_1 = 0.6943$ μm and $\lambda_2 = 0.3471$ μm, that is, at the fundamental and the first harmonic wavelength.

The *hook method* that combines an optical interferometer with a spectrograph is another interferometric technique for determining densities in high temperature gases (see, e.g., Huber, 1971; van de Weijer and Cremers, 1983). After having passed through the test field, the interferometric light beam is focused onto the entrance slit of a spectrograph. The interference fringes appear superimposed on the spectrum. In a usual arrangement, the recorded spectrogram shows parallel, oblique fringes. If the test gas, at the high temperatures, has absorption lines, the fringes exhibit a hook on each side of such line, and the fringe pattern in between the two

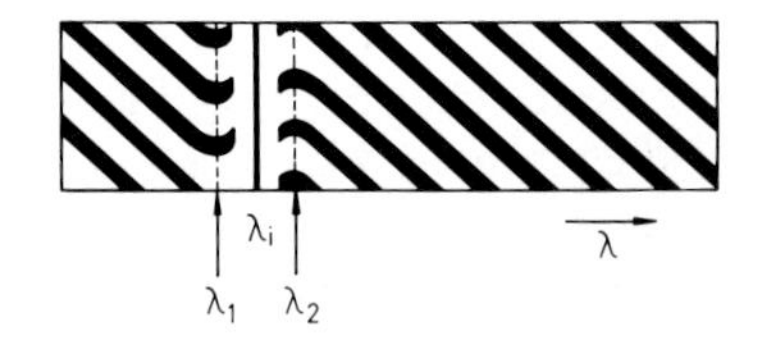

Fig. 3.62 Schematic representation of a hook interferogram with the resonance wavelength λ_i and hooks at positions λ_1 and λ_2.

hooks corresponds to the anomalous dispersion near absorption (Fig. 3.62; see also Fig. 3.1). The formation of a hook is described by

$$b \frac{d(n-1)}{d\lambda} = K_{\mathrm{H}} \tag{3.57}$$

where b is the width of the test section (in the z direction), and n and λ are the refractive index and the wavelength, respectively. The "hook constant" K_{H} depends only on the data of the apparatus and must be determined by calibration. By inserting the quantum mechanical expression for n into Eq. (3.57), one obtains a quadratic equation in λ. The two roots of this equation, λ_1 and λ_2, describe the positions of the hooks in the spectral interferogram. Conversely, by measuring λ_1, λ_2, and the resonant wavelength λ_i, one obtains an expression for the product $(N_i f_i)$, where f_i is the oscillator strength (see Eq. 3.5), and N_i the number density of the gas atoms or molecules in the lower state of the respective electron transition. With a catalogue of f_i values available, one may determine the gas density ρ.

3.4. Phase Contrast and Field Absorption

In Section 3.2.2.3 we have discussed devices that affect the intensity of the light in the focal plane of the schlieren head (see, e.g., Fig. 3.17). A diffractive image of the (pointlike) light source is formed in this plane; it consists of the central zeroth order of circular shape, and of the annular diffractive images of higher order. In this section, methods will be described with which the diffraction pattern in the focal plane of the schlieren head is somehow manipulated. They are discussed here separately, because their working principles cannot be analyzed by geometrical ray theory. These methods do not play a dominant role in optical flow visualization, but some of them can be very useful for investigating flow fields with weak density differences.

A consequent step in the development of highly sensitive visualization methods was the adaption of Zernike's phase contrast method to the study of gas flows. The application of this method requires the use of a coherent light source and an optical setup similar to that of a schlieren system (Fig. 3.11). In the focal point of the schlieren head a minute phase plate that coincides with the zeroth-order diffractive image changes the phase of the zeroth-order light by an angle of 90°. The principle of the phase contrast method is explained with the aid of a vector diagram (Fig. 3.63). This diagram shows both amplitude and direction or phase for light rays or waves having passed through the test field. The radial vector with phase angle $\varphi = 0$ designates the incident, undisturbed light. The phase shift experienced by the light in the test field is expressed by a certain value of the phase angle φ, that describes at the same time the phase difference between the disturbed and the undisturbed (incident) ray. Since the test field is transparent, the amplitudes of all transmitted rays are equal, and the end points of the respective light vectors are found on a circle of a radius that is equal to the amplitude of the incident light. The principal role of the phase contrast method is to visualize weak density differences, and the disturbed light is therefore assumed to be changed by small phase angles φ_i. The vectors of three disturbed light rays or waves—1, 2, and 3—with the respective phase angles φ_1, φ_2, and φ_3 are shown in Fig. 3.63. It would be impossible to detect the phase differences between these waves from an ordinary two-beam interferogram, and owing to the constant length of these light vectors, the illumination in the respective portion of the recording plane would be almost uniform.

One may separate each light vector in Fig. 3.63 into two components: one common to all light vectors (here 1, 2, and 3) and a second, individual component. The end point A of the common component represents the

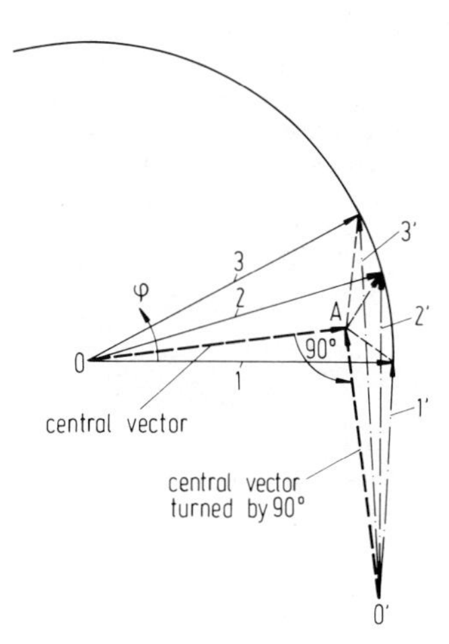

Fig. 3.63 Vector diagram representing the principles of the phase contrast method. Notation is explained in the text.

center of gravity of the area, here: the triangle, formed by the end points of the considered light vectors; and for determining A, each light vector is "weighted" according to the portion of the field of view illuminated by the light of the respective vector. The light associated with the common vector component represents the light of the zeroth diffraction order in the focal point of the schlieren head. An obvious explanation for this is that the intensity of the zeroth-order light is constant throughout the field of view, independent of the light's phase angle. The individual vector components are then equivalent to the light of the higher diffractive orders. With the aid of the phase plate mentioned earlier, the phase of the common vector component is turned by 90°. The result is shown in Fig. 3.63: The new vectors, denoted by 1′, 2′, and 3′, are now of different lengths. As a consequence, the intensity in the field of view is variable and changes as function of the phase angle. The optical phase differences are thereby transformed into alterations of amplitude or contrast (Fig. 3.64).

Since the test field had been assumed to generate only weak phase differences, the length of the common vector component ("central vec-

Fig. 3.64 Supersonic flow at Mach number $M_\infty = 3.74$ over a two-dimensional profile as visualized by the phase-contrast method without absorption. (From Philbert, 1964.)

tor") is not much smaller than the radius of the circle whose radius is equal to the length of the undisturbed light vector ($\varphi = 0$). If this radius is assumed to be one, the lengths of the new vectors, after having turned the central vector by 90°, are approximately given by $1 + \varphi_i$, where the φ_i represent the respective phase angles. Since φ_i is a small quantity, the intensity of the light associated to the new vectors is $1 + 2\varphi_i$; and the contrast in the field of view or in the recording plane is $2\varphi_i$. This contrast can be increased if one provides a certain degree of absorption to the phase plate, so that the length of the turned central vector is decreased to a value α, with $0 < \alpha < 1$. The lengths of the new vectors are now $\alpha + \varphi_i$, and the contrast has been increased to a value $2\varphi_i/\alpha$. The sensitivity of the phase contrast method, or the ability to resolve small density differences, is estimated to exceed the respective sensitivity of interferometric methods by a factor of ten (Philbert, 1964; Véret, 1970). If the phase change throughout the field of view is more than 360°, fringes appear like in an ordinary interferogram, and the evaluation is equivalent to that for two-beam interferometry. In a practical setup, Anderson and Taylor (1982) used a phase plate of 0.2 mm in diameter evaporated on fused silika, while their test beam had an open diameter of 50 mm.

A number of modifications of the schlieren system are known, in which a certain portion of the light source image in the focal plane of the schlieren head is shifted in phase. A shift of one-half of this light by an angle of 180°, realized by a respective phase plate, results in a dark field of view due to the interference of the light having passed through the phase plate with the unshifted light. Refractive index variations in the test field will appear as bright disturbances in the dark field. Such "dephasing schlieren" systems, which have been reviewed by Wolter (1956), have been developed mainly for being used in the microscopy of phase objects, but they are of little importance for flow visualization.

The previously used vector diagram is very appropriate for explaining the principle of an interferometric method developed by Erdmann (1951). The setup is again that of a schlieren system, and the method which is not restricted to small phase changes uses a particular absorption plate in the focal plane of the schlieren head. While the light of the zeroth diffraction order remains unchanged, the intensity of the higher orders is reduced by appropriate absorption. For this purpose, the absorption plate placed in the focal point of the schlieren head has a central aperture of size equal to that of the zeroth diffraction order image. The result of this optical procedure is shown in the vector diagram (Fig. 3.65) for three light vectors 1, 2, 3 of different phase angle but equal length or amplitude. The relative increase of the intensity of the zeroth diffraction order is equivalent to a lengthening of the central vector whose end point is in A. The end points

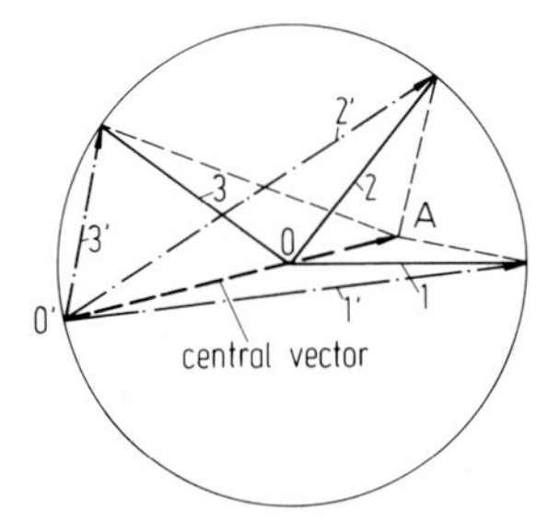

Fig. 3.65 Vector representation of Erdmann's field absorption method. Original light vectors are denoted by 1, 2, and 3; new vectors after lengthening of the central vector are 1′, 2′, and 3′.

of the three light vectors remain unchanged because the optical phase is not altered with this arrangement. But with the new origin 0′, the new vectors 1′, 2′, and 3′ have different lengths. A maximum of contrast is achieved if the absorption is chosen such that the new origin 0′ falls onto the phase circle. Then, the intensity of the light associated with the new vectors varies between zero and 2^2, if the radius of the phase circle is 1. It is obvious that the phase angle is allowed to vary by more than 360°. Starting with the new origin, where the respective intensity is zero, the intensity will again be zero after every full period of 360°. Hence, fringes are obtained that are identical with those known for ordinary two-beam interferometry.

Erdmann (1951) has used this "field absorption method" for visualizing compressible flow fields in a supersonic wind tunnel. Peterka and Richardson (1984) employ a system, which is analoguous to Erdmann's method: a thin wire in the focal plane of the schlieren head substracts the zeroth diffraction order. The test field is placed in only one half of the parallel schlieren light. After the subtraction of the zeroth order, interference (of the higher orders) of the disturbed and the undisturbed light is allowed, thus resulting in a fringe pattern like in a reference beam interferometer.

3.5. Streaming Birefringence

A number of liquids or liquid solutions exhibit the optical effect of streaming birefringence, that is, these liquids become birefringent under the action of shear forces in a flow. Birefringence is also known for a number of transparent solid materials, mainly crystals. The light propagation in such a material is directionally dependent. An incident light wave is separated in the birefringent medium into two components, which are linearly polarized, with the planes of polarization being perpendicular to one another, and which propagate at different phase velocities. Therefore, different values of the refractive index are assigned to the two components. They are out of phase when leaving the birefringent material, and this difference in optical phase can be visualized and measured by interferometric means. Many attempts have been made for using the effect of streaming birefringence, known for more than one century (Mach, 1873; Maxwell, 1873), for determining shear rates or deformation velocities, or just for the visualization of the flow of such liquids.

While birefringence is a permanent effect in crystals, it occurs in the respective liquids only when they are subjected to shear forces (i.e., when they flow). Temporary birefringence can also be generated in a number of solid, noncrystalline materials by applying appropriate stresses. Making use of this effect of "photoelasticity" has led to the development of the methods of optical stress analysis in solid mechanics (see, e.g., Kuske and Robertson, 1974). Many of the techniques applied to streaming birefringes have been adopted from respective methods of photoelasticity.

Birefringence can be generated in fluids consisting of elongated and deformable molecules (polymers) or having elongated, solid, crystal-like particles in solution (coloidals). If the fluid is at rest, these particles or molecules are randomly distributed, and the fluid is optically isotropic, and thus not birefringent. The shear forces in a flow cause the particles or long molecules to align in a preferential direction, the fluid is then anisotropic or birefringent. Theoretically, any fluid consisting of nonspherical particles or molecules should show this effect. Boyer *et al.* (1978) verified experimentally that the effect, though weak, even exists in air flows with strong shear. For the purpose of flow visualization or measurement, however one is interested in high optical sensitivity, and this restricts our discussion to such fluids, of which the particles (or molecules) are of a much larger dimension than the molecules of simple or normal substances.

As stated in the beginning, flow visualization by streaming birefringence is based on the refractive index behaviour of the fluid, and optical interference is the principal technique for visualization. For the purpose of possible quantitative evaluation one needs a relationship between the observable refractive index field and the state of the flow, a so-called flow-optic relation. The respective theory is not totally developed, and an evaluation with the aim of determining the flow quantities often generates difficult numerical problems. The respective theories have not been developed primarily for performing flow measurements. Instead, the interest was to determine from a simple flow pattern characteristic molecular constants or physico-chemical properties of the fluid. The double refraction in polymers has been analyzed by Philippoff (1964), while Wayland (1964) investigated the effect of streaming birefringence in colloidal solutions.

An aqueous solution of Milling yellow dye (chemical nomenclature: H5G) is known to have high optical sensitivity and has been used therefore in many visualization experiments. This commercial dye consists of the crystals of the pure dye and certain additives (e.g., Na_2SO_4 and NaCl). Solving the powdered dye in water requires the addition of energy, for example, in the form of heat or ultrasound waves. The dye crystals, about 1–2 μm long, are birefringent. When the crystals suspended in the solution align under the action of flow shear, the gross effect is that the whole solution becomes birefringent to a certain degree. The properties of a Milling yellow solution have been described by Swanson and Green (1969) and by Pindera and Krishnamurthy (1978a). This colloidal liquid exhibits strong non-Newtonian behavior, and only for very low deformation velocities can the liquid be regarded to behave like a Newtonian fluid. The viscous behavior depends sensitively on temperature and dye concentration. Schmitz and Merzkirch (1984) have shown that the flow curve of a particular concentration of Milling Yellow is, for a wide range of deformation velocities, exactly proportional to the flow curve of human blood (Fig. 3.66).

The oldest and most widely used technique for visualizing the flow of a birefringent liquid is an apparatus called polariscope that is also employed for photoelastic experiments (Fig. 3.67). A linearly polarized beam of light

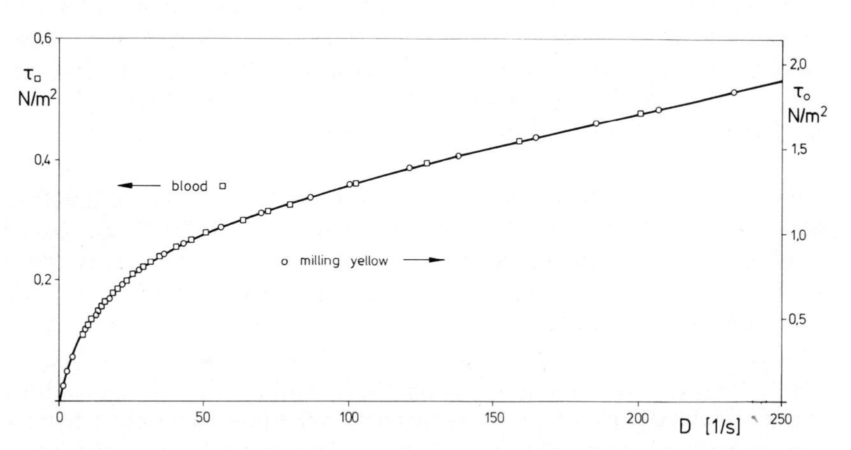

Fig. 3.66 Flow curve of an aqueous solution of Milling yellow dye (0.26% pure dye, 0.56% Na_2SO_4, 0.13% NaCl) and of 50% hematocrit human blood. Note the different scales for the shear stress τ. (From Schmitz and Merzkirch, 1984.)

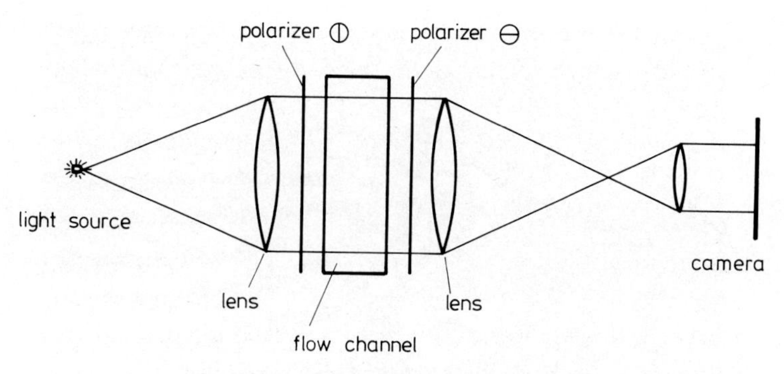

Fig. 3.67 Optical setup of a polariscope.

is directed through the flow, and after having traversed a second polarizer ("analyzer"), the two beams resulting from the light separation due to double refraction can interfere with one another. In most cases it can be assumed that the optical axis of the birefringent fluid is perpendicular to the direction of light propagation (the z direction). Then, the two beams are not separated in space, they rather coincide, but the light waves propagate at different speed in the fluid, and they are out of phase when leaving the test field. The optical phase difference experienced by two interfering light rays in a two-dimensional test field of constant thickness b (in z direction) is

$$\Delta\varphi/2\pi = \Delta n(b/\lambda),$$

where λ is the light wave length, and $\Delta n(x, y)$ is the difference of the two refractive indices assigned to the two separated rays. Fringes ("isochromates") appear in the field of view at positions where the quantity $\Delta\varphi/2\pi$ is equal to an integer. The isochromatic fringe pattern (Fig. 3.68) yields the distribution of the experimental data $\Delta n(x, y)$.

Wayland (1974) has derived a flow-optic relation that expresses Δn as a power series of the quantity $(\dot{\varepsilon}_{max}/D)$, where $\dot{\varepsilon}_{max}$ is the maximum deformation velocity of a plane (two-dimensional) flow, and D is the coefficient of rotary diffusion counteracting the aligning mechanism of the shear forces. Attempts for quantitative evaluation have been made for only accounting for the linear term of the power series:

$$\Delta n(x, y)(b/\lambda) = a(\dot{\varepsilon}_{max}/D) \tag{3.58}$$

where a is a constant that has to be determined by calibration. This linear approximation restricts the validity of Eq. (3.58) to small values of $\dot{\varepsilon}_{max}$, or to very low flow velocities or low Reynolds numbers (creeping flow). In this range, the flow curve of the Milling yellow solution (Fig. 3.66) is

Fig. 3.68 Isochromatic fringe pattern of the flow over a step configuration. (From E. Schmitz and W. Merzkirch, Universität Essen.)

almost linear, and it follows that Eq. (3.58) applies to flow conditions under which this liquid behaves like a Newtonian fluid.

The quantity $\dot{\varepsilon}_{max}$ in Eq. (3.58) is related to the velocity components u, v of the plane flow by

$$\dot{\varepsilon}^2_{max} = 4\left(\frac{\partial u}{\partial x}\right)^2 + \left(\frac{\partial u}{\partial y} + \frac{\partial v}{\partial x}\right)^2. \tag{3.59}$$

For the purpose of a quantitative evaluation of the velocity field, Peebles and Liu (1965) have introduced a stream function $\psi(x,y)$, and Eq. (3.58) then becomes

$$\left(2\frac{\partial^2\psi}{\partial x\,\partial y}\right)^2 + \left(\frac{\partial^2\psi}{\partial y^2} - \frac{\partial^2\psi}{\partial x^2}\right)^2 = \frac{Db}{a\lambda}\,\Delta n(x, y). \tag{3.60}$$

On the right-hand side one has the experimentally determined data. Since the measured data of Δn is available only along the isochromatic fringes, one has to provide an appropriate interpolation for obtaining a continuous pattern $\Delta n(x, y)$. The integration of the highly nonlinear Eq. (3.60) generates enormous numerical difficulties, and it requires the knowledge of some boundary values of ψ (e.g., along a line of symmetry or at a solid wall). Solutions for u and v in the flow through channels of varying cross sections, including step-wise expansions, have been found with satisfactory accuracy by Prados and Peebles (1959), Peebles and Liu (1965), and Horsmann *et al.* (1979), while Durelli and Norgard (1972) combined the method with the hydrogen bubble technique.

Another problem arising from the linear approximation expressed by Eq. (3.58) is that, at the very low velocities for which Eq. (3.58) holds, the number of available isochromatic fringes in the field is limited (i.e., the data density is low). The number of fringes increases with increasing mean velocity, but then Eq. (3.58) might be violated, and at the same time, the non-Newtonian behavior of the fluid might no longer be negligible. It appears that the interest in the method for quantitative use is in the very low velocity range, where other measuring techniques fail to produce reliable results. The pattern visualized with the polariscope may include a second, different fringe system ("isoclines"), which carries additional information. The data density of the isoclinic fringes, however, is even much lower than that of the isochromates, and it has not been used for flow measuring purposes.

A technique that enables one to use streaming birefringence without being restricted by the amount of the flow velocity, nor by the viscous behavior of the fluid has been described by Schmitz and Merzkirch (1983). The two refractive indices, n_1 and n_2, are measured separately by means of a Mach–Zehnder interferometer (Fig. 3.69). Use is made of the fact

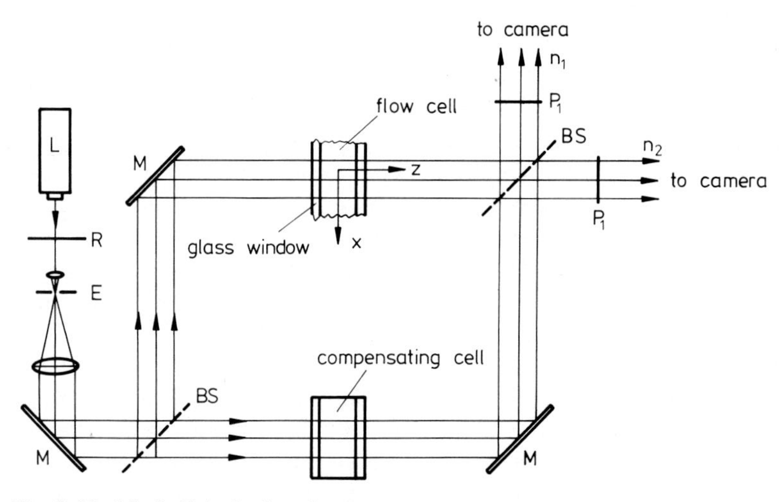

Fig. 3.69 Mach–Zehnder interferometer applied to streaming birefringence. The compensating cell is filled with the birefringent fluid at rest. The two exits after the second beam splitter (BS) are used for quantitative evaluation. Directions of polarization of the two polarizers P_1 are perpendicular to each other.

that this interferometer has two exits (see Section 3.3.2.1), and that the two waves separated due to double refraction have different linear polarization. In each of the two exits one has to provide a polarizer with a direction of polarization appropriate for passing only one of the two waves. This separate measurement of n_1 and n_2 delivers more information than the classical polariscope by which one determines only the difference $\Delta n = n_1 - n_2$. The quantitative evaluation of the measured data requires the existence of equations relating the refractive indices, and not their difference, to the state of the flow. Schmitz and Merzkirch (1983) develop such a theory on the basis of the early work of Boeder (1932), who had neglected the finite size of the colloidal particles causing the birefringence. The procedure requires integrating a nonlinear equation of the same type as Eq. (3.59). The integration and thereby the quantitative evaluation has been achieved for nearly Newtonian but also for strongly non-Newtonian flow between plane parallel plates.

In the Introduction it had been stated that scattered light methods may provide local information on three-dimensional flow fields. Several attempts have been made for applying the scattered light technique to streaming birefringence. It is known that the light scattered from a birefringent medium exhibits a characteristic pattern from which one may conclude onto the state of birefringence. Like in other applications it is assumed that the pattern observable in the scattered radiation is caused by the state of the flow at the position, where the light is scattered from, and that this radiation is not affected during its passage through the fluid. Pindera and Krishnamurthy (1978b) investigated some fundamental optical parameters of a Milling yellow solution in scattered light, while McAfee and Pih (1971, 1974) demonstrated the possibility of determining quantities of the flow through simple geometries from the observation of scattered light. A theoretically derived flow-optic relation necessary for the quantitative evaluation is not available for streaming birefringent light scattering. McAfee and Pih (1974) therefore proposed an empirical relationship that was tested for relatively simple flow cases. Horsmann and Merzkirch (1981) extended these investigations and developed, on an empirical basis, a flow-optic relationship that applies to the general three-dimensional case.

Three thin, plane sheets of laser light (see Fig. 2.12) are directed in three orthogonal directions through the flow field. In the respective fields of the scattered radiation one may observe patterns of fringes whose formation is explained by optical interference (Fig. 3.70). The spacing between the fringes is not uniform, so that one can define gradients $\partial N/\partial s$, where N designates the order number of a fringe, and s is the direction of

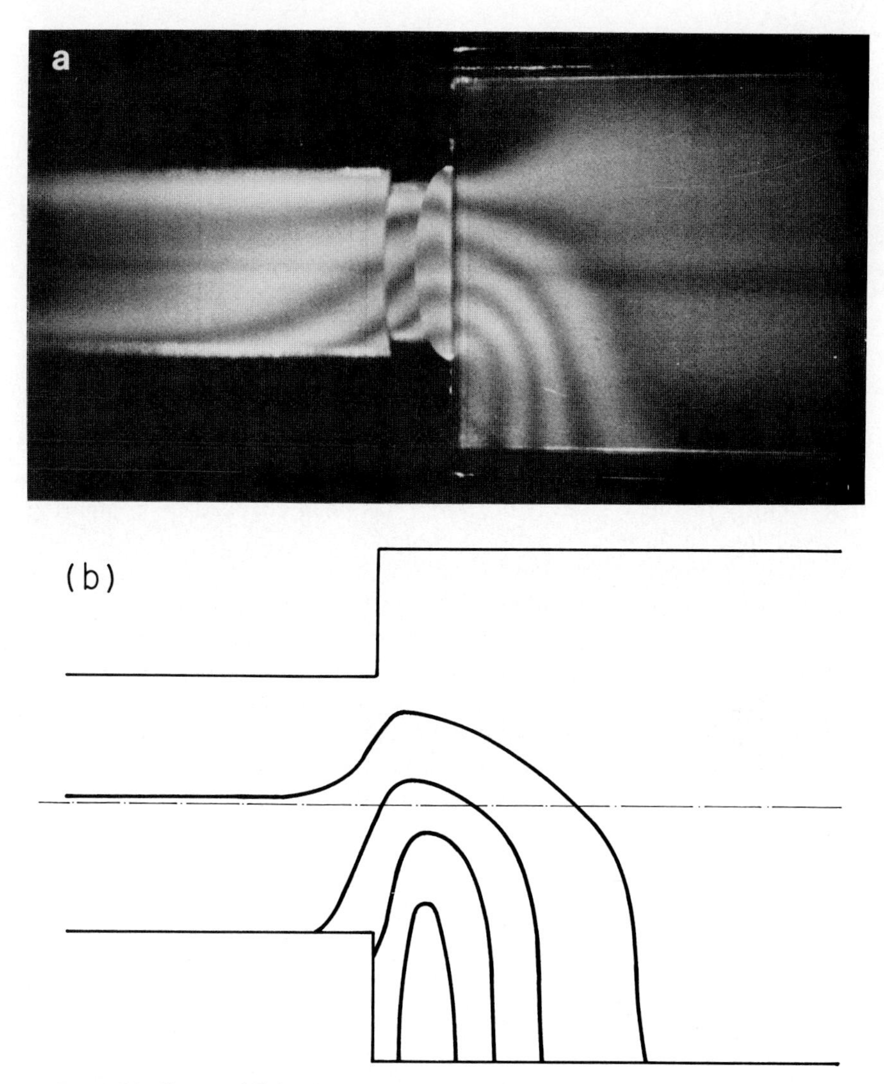

Fig. 3.70 Scattered light photograph (a) of the flow in a circular pipe with a step-wise increase of the cross-section, and (b) fringe pattern calculated by using Eq. (3.61). (From Horsmann and Merzkirch, 1981.)

the incident light in each sheet (x, y, or z). With the definition of maximum deformation velocities $\dot{\varepsilon}_{\max}$ for each plane xy, yx, zx (see index), analoguous to Eq. (3.59), the relationships derived by Horsmann and Merzkirch (1981) are

$$\frac{\partial N}{\partial z} = \{C_1^2\dot{\varepsilon}^2_{\max_{xy}} + C_2^2(\dot{\varepsilon}^2_{\max_{yz}} + \dot{\varepsilon}^2_{\max_{zx}})\}^{1/2},$$

$$\frac{\partial N}{\partial x} = \{C_1^2\dot{\varepsilon}^2_{\max_{yz}} + C_2^2(\dot{\varepsilon}^2_{\max_{zx}} + \dot{\varepsilon}^2_{\max_{xy}})\}^{1/2}, \tag{3.61}$$

$$\frac{\partial N}{\partial y} = \{C_1^2\dot{\varepsilon}^2_{\max_{zx}} + C_2^2(\dot{\varepsilon}^2_{\max_{xy}} + \dot{\varepsilon}^2_{\max_{yz}})\}^{1/2}.$$

This system includes two constants, C_1 and C_2, that have to be determined by calibration. C_2, which is an order of magnitude smaller than C_1, applies to the contribution of the shear rates in the two planes including the direction of the incident light. The constants depend on temperature, concentration of Milling yellow dye in the solution, and on the wavelength of the used light, similarly to the behavior of the constant a in Eq. (3.58). The validity of the system [Eq. (3.61)] has been checked by comparing visible fringe patterns with patterns computed by using the Eq. (3.61), see Fig. 3.70. A direct evaluation for determining the velocity, however, is again associated with a number of immense numerical difficulties.

In the reported experiments, that have been performed with either one of the three optical techniques (polariscope, Mach–Zehnder interferometer, scattered light technique), an aqueous solution of Milling yellow dye was used. The use of other birefringent fluids for visualization experiments has been reported, e.g., polyvinyl alcohol (Nakatani *et al.*, 1982), and vanadium pentoxide (Liepsch *et al.*, 1982), but none of them is as sensitive as Milling yellow.

References

Section 3.1

Alpher, R. A., and White, D. R. (1959a). Optical refractivity of high temperature gasses. I: Effects resulting from dissociation of diatomic gases. *Phys. Fluids* **2,** 153–161.

Alpher, R. A., and White, D. R. (1959b). Optical refractivity of high temperature gases. II: Effects resulting from ionization of monatomic gases. *Phys. Fluids* **2,** 162–169.

Anderson, J. H. B. (1969). Experimental determination of the Gladstone-Dale constants for dissociating oxygen. *Phys. Fluids* **12,** Suppl., I-57–I-60.

Barnes, N. F. (1953). Optical techniques for fluid flow. *J. SMPTE* **61,** 136–160.

Bershader, D. (1971). Some aspects of the refractive behavior of gases. *In* "Modern Optical

Methods in Gas Dynamic Research" (D. S. Dosanjh, ed.), pp. 65–83. Plenum, New York.
Blendstrup, G., Bershader, D., and Langhoff, P. W. (1978). Resonance refractivity studies of sodium vapor for enhanced flow visualization. *AIAA J.* **16,** 1106–1108.
Burner, A. W., and Goad, W. K. (1980). Gladstone–Dale constant for CF_4. *NASA Tech. Memo.* **NASA TM-X-80228.**
Dobbins, H. M., and Peck, E. R. (1973). Change of refractive index of water as a function of temperature. *J. Opt. Soc. Am.* **63,** 318–320.
Fiedler, H., Nottmeyer, K., Wegener, P. P., and Raghu, S. (1985). Schlieren photography of water flow. *Exp. Fluids* **3,** 145–151.
Goldstein, R. J. (1976). Optical measurement of temperature. *In* "Measurement in Heat Transfer" (E. R. G., Eckert and R. J. Goldstein, eds.), pp. 241–293. Hemisphere, Washington, D.C.
Goldstein, R. J. (1983). Optical systems for flow measurement: shadowgraph, schlieren, and interferometric techniques. *In* "Fluid Mechanics Measurements" (R. J. Goldstein, ed.), pp. 377–422. Hemisphere, Washington, D.C.
Grange, B. W., Stevenson, W. H., and Viskanta, R. (1976). Refractive index of liquid solutions at low temperatures: an accurate measurement. *Appl. Opt.* **15,** 858–859.
Hauf, W., and Grigull, U. (1970). Optical methods in heat transfer. *Adv. Heat Transfer* **6,** 131–366.
Jones, F. E. (1980). Simplified equation for calculating the refractivity of air. *Appl. Opt.* **19,** 4129–4130.
Kügler, E., and Bershader, D. (1983). Recent high-resolution resonant refractivity studies of a sodium-seeded flame. *Exp. Fluids* **1,** 51–55.
Lauterborn, W., and Vogel, A. (1984). Modern optical techniques in fluid mechanics. *Annu. Rev. Fluid Mech.* **16,** 223–244.
Lensch, G. (1977). Rotationsrelaxation von Wasserstoff und Deuterium im Kryo-Stosswellenrohr. Dissertation, RWTH, Aachen.
Merzkirch, W. (1981). Density-sensitive flow visualization. *In* "Fluid Dynamics (R. J. Emrich, ed.), Methods in Experimental Physics" Vol. 18A, pp. 345–403. Academic Press, New York.
Peters, F. (1985). Schlieren interferometry applied to a gravity wave in a density-stratified liquid. *Exp. Fluids* **3,** 261–269.
Stirnberg, D. (1982). Interferometrische Konzentrations- und Temperaturmessungen in einer laminaren Diffusionsflamme. Dissertation, Ruhr-Universität, Bochum.
Weinberg, F. J. (1963). "Optics of Flames." Butterworth, London.
Wettlaufer, D. E., and Glass, I. I. (1972). Specific refractivities of atomic nitrogen and oxygen. *Phys. Fluids* **15,** 2065–2066.
Weyl, F. J. (1954). Analysis of optical methods. *In* "Physical Measurements in Gas Dynamics and Combustion" (R. W. Ladenburg, ed.), pp. 3–25. Princeton Univ. Press, Princeton, New Jersey.
Witte, A. B., and Collins, D. J. (1971). Basics of flow visualization. *Fundam. Aerosp. Instrum.* **4,** 37–64.

Section 3.2

Andrews, J. R., and Netzer, D. W. (1976). Laser schlieren for study of solid-propellant deflagration. *AIAA J.* **14,** 410–412.

Bander, J. A., and Sanzone, G. (1974). An improved laser-schlieren system for the measurement of shock-wave velocity. *Rev. Sci. Instrum.* **45,** 949–951.

Bar-Ziv, E., Sgulim, S., Kafri, O., and Keren, E. (1983). Temperature mapping in flames by moiré deflectometry. *Appl. Opt.* **22,** 698–705.

Bathelt, A. G., Viskanta, R., and Leidenfrost, W. (1979). An experimental investigation of natural convection in the melted region around a heated horizontal cylinder. *J. Fluid Mech.* **90,** 227–239.

Beams, J. W. (1954). Shadow and schlieren methods. *In* "Physical Measurements in Gas Dynamics and Combustion" (R. W. Ladenburg, ed.), pp. 26–46. Princeton Univ. Press, Princeton, New Jersey.

Buzzard, R. D. (1968). Description of three-dimensional schlieren system. *Proc. Int. Congr. High-Speed Photogr., 8th,* pp. 335–340.

Carey, V. P., and Mollendorf, J. C. (1978). Measured variation of thermal boundary-layer thickness with Prandtl number for laminar natural convection from a vertical uniform-heat-flux surface. *Int. J. Heat Mass Transfer* **21,** 481–488.

Chashechkin, J. D. (1985). Colour schlieren method. *In* "Optical Methods in Dynamics of Fluids and Solids" (M. Pichal, ed.), pp. 275–282. Springer-Verlag, Berlin and New York.

Corcoran, J. W. (1967). Applications of the Isodensitytracer in high-speed photography. *Kurzzeitphotogr., Ber. Int. Kongr. Kurzzeitphotogr. Hochfrequenzkinematogr., 7th, Zurich, 1965,* pp. 466–471.

Davis, M. R. (1972). Quantitative schlieren measurements in a supersonic turbulent jet. *J. Fluid Mech.* **51,** 435–447.

Davis, M. R. (1982). Coherence between large-scale jet mixing structure and its pressure field. *J. Fluid Mech.* **116,** 31–57.

Debrus, S., Françon, M., Grover, C. P., May, M., and Roblin, M. L. (1972). Ground glass differential interferometer. *Appl. Opt.* **11,** 853–857.

Decker, G., Deutsch, R., Kies, W., and Rybach, J. (1985). Computer-simulated schlieren-optics. *Appl. Opt.* **24,** 823–828.

Dixon, W. P. (1982). Radial schlieren for detecting small index gradients. *Appl. Opt.* **21,** 1896.

Dvorak, V. (1880). Über eine neue einfache Art der Schlierenbeobachtung. *Ann. Phys. Chem.* **9,** 502–512.

Farrell, P. V., and Hofeldt, D. L. (1984). Temperature measurement in gases using speckle photography. *Appl. Opt.* **23,** 1055–1059.

Fisher, M. J., and Krause, F. R. (1967). The crossed-beam correlation technique. *J. Fluid Mech.* **28,** 705–717.

Foucault, L. (1859). *Ann. Obs. (Paris)* **5,** 197.

Gärtner, U. (1983). Visualization of particle displacement and flow in stratified salt water. *Exp. Fluids* **1,** 55–56.

Grandke, T. (1985). Theory and application of the laser shadow technique. Part 1: Two-dimensional turbulent flows. *Exp. Fluids* **3,** 77–85.

Grossin, R., Jannot, M., and Viannay, S. (1971). Schlieren visualization device allowing an arbitrary orientation of the lines with respect to the scanning direction. *Appl. Opt.* **10,** 201–204.

Heavens, S. N. (1980). Visualization of the acoustic excitation of a subsonic jet. *J. Fluid Mech.* **100,** 185–192.

Hesselink, L., and White, B. S. (1983). Digital image processing of flow visualized photographs. *Appl. Opt.* **22,** 1454–1461.

Holder, D. W., and North, R. J. (1952). A schlieren apparatus giving an image in colour. *Nature (London)* **169,** 466.

Holder, D. W., and North, R. J. (1963). "Schlieren Methods," Notes on Applied Science, No. 31. HM Stationary Off., London.

Holder, D. W., North, R. J., and Wood, G. P. (1956). Optical methods for examining the flow in high speed wind tunnels. *AGARDOgraph* **AGARD-AG-23.**

Hosch, J. W., and Walters, J. P. (1977). High spatial resolution schlieren photography. *Appl. Opt.* **16,** 473–482.

Houtman, H., and Meyer, J. (1984). Multipass dark-ground photography of a low-pressure gas jet. *Appl. Opt.* **23,** 2178–2180.

Howes, W. L. (1984). Rainbow schlieren and its applications. *Appl. Opt.* **23,** 2449–2460.

Hsia, Y., Baganoff, D., Krothapalli, A., and Karamcheti, K. (1984). An enhanced flow visualization technique for planar free shear layers. *AIAA J.* **22,** 439–441.

Ishikawa, N. (1983). Experimental study of jet mixing mechanisms in a model secondary combustor. *AIAA J.* **21,** 565–571.

Kafri, O. (1981). Noncoherent method for mapping phase objects. *Opt. Lett.* **5,** 555–557.

Kafri, O., Livnat, A., and Glatt, I. (1984). Temporally stable density patterns in liquids. *J. Fluids Eng.* **106,** 257–261.

Kantrowitz, A., and Trimpi, R. L. (1950). A sharp-focusing schlieren system. *J. Aeronaut. Sci.* **17,** 311–314.

Kent, J. C. (1969). Fabrication of graded filters for knife-edge replacement in laser schlieren optical systems. *Appl. Opt.* **8,** 2148–2149.

Keren, E., Bar-Ziv, E., Glatt, I., and Kafri, O. (1981). Measurement of temperature distribution of flames by moiré deflectometry. *Appl. Opt.* **20,** 4263–4266.

Kessler, T. J., and Hill, W. G. (1966). Schlieren analysis goes to color. *Aeronat. Astronaut.* **4,** 38–40.

Kiefer, J. H., and Lutz, R. W. (1965). Simple quantitative schlieren technique of high sensitivity for shock tube densitometry. *Phys. Fluids* **8,** 1393–1394.

Kiefer, J. H., Al-Alami, M. Z., and Hajduk, J.-C. (1981). Physical optics of the laser schlieren shock tube technique. *Appl. Opt.* **20,** 221–230.

Klein, E. J. (1970). A planview shadowgraph technique for boundary layer visualization. *AIAA J.* **8,** 963–965.

Knöös, S. (1968). A quantitative schlieren technique for measuring one-dimensional density gradients in transparent media. *Proc. Int. Congr. High-Speed Photogr., 8th,* pp. 346–350.

Köpf, U. (1972). Application of speckling for measuring the deflection of laser light by phase objects. *Opt. Commun.* **5,** 347–350.

Kogelschatz, U., and Schneider, W. R. (1972). Quantitative schlieren techniques applied to high current arc investigations. *Appl. Opt.* **11,** 1822–1832.

Koziak, W. W. (1970). Quantitative space and time resolved laser schlieren system for the study of hypersonic flow. *Rev. Sci. Instrum.* **41,** 1770–1773.

Larmore, L., and Hall, F. F., Jr. (1971). Optics for the airborne observer. *J. Soc. Photo-Opt. Instrum. Eng.* **9,** 87–94.

Maddox, A. R., and Binder, R. C. (1971). A new dimension in the schlieren technique: Flow field analysis using color. *Appl. Opt.* **10,** 474–481.

McEwan, A. D. (1983). The kinematics of stratified mixing through internal wave breaking. *J. Fluid Mech.* **128,** 47–57.

Merzkirch, W., and Erdmann, W. (1974). Measurement of shock wave velocity using the Doppler principle. *Appl. Phys.* **4,** 363–366.

Mowbray, D. E. (1967). The use of schlieren and shadowgraph techniques in the study of flow patterns in density stratified liquids. *J. Fluid Mech.* **27,** 595–608.
North, R. J., and Stuart, C. M. (1963). Flow visualization and high-speed photography in hypersonic aerodynamics. *Proc. Int. Congr. High-Speed Photogr., 6th,* pp. 470–477.
O'Hare, J. E., and Trolinger, J. D. (1969). Holographic color schlieren. *Appl. Opt.* **8,** 2047–2050.
Pfeifer, H. J., vom Stein, H. D., and Koch, B. (1970). Mathematical and experimental analysis of light diffraction on plane shock waves. *Proc. Int. Congr. High-Speed Photogr., 9th* pp. 423–426.
Philbert, M. (1964). Visualisation des écoulements à basse pression. *Rech. Aerosp.* No. 99, 39–48.
Prescott, R., and Gayhart, E. L. (1951). A method of correction of astigmatism in schlieren systems. *J. Aerosp. Sci.* **18,** 69.
Roos, F. W., and Bogar, T. J. (1982). Comparison of hot film probe and optical techniques for sensing shock motion. *AIAA J.* **20,** 1071–1076.
Rotem, Z., Hauptmann, E. G., and Claassen, L. (1969). Semifocusing color schlieren system for use in fluid mechanics and heat transfer. *Appl. Opt.* **8,** 2327–2328.
Royer, H., and Smigielski, P. (1968). Méthodes strioscopiques appliquées à l'étude des sillages hypersoniques à basse pression. *Proc. Int. Congr. High-Speed Photogr., 8th,* pp. 359–361.
Sajben, M., and Crites, R. C. (1979). Real-time optical measurement of time-dependent shock position. *AIAA J.* **17,** 910–912.
Schardin, H. (1934). Das Toeplersche Schlierenverfahren. *VDI—Forschungsh.* No. 367.
Schardin, H. (1942). Die Schlierenverfahren und ihre Anwendungen. *Ergeb. Exakten Naturwiss.* **20,** 303–439.
Schardin, H. (1958). Ein Beispiel zur Verwendung des Stosswellenrohres für Probleme der instationären Gasdynamik. *Z. Angew. Math. Phys.* **9b,** 606–621.
Schmidt, M. C., and Settles, G. S. (1986). Alignment and application of the conical shadowgraph flow visualization technique. *Exp. Fluids* **4,** 89–92.
Schwar, M. J. R., and Weinberg, F. J. (1969). The measurement of velocity by applying schlieren interferometry to Doppler shifted laser light. *Proc. R. Soc. London, Ser. A* 311, 469–476.
Sedney, R. (1972). Visualization of boundary layer flow patterns around protuberances using an optical-surface indicator technique. *Phys. Fluids* **15,** 2439–2441.
Settles, G. S. (1970). A direction-indicating color schlieren system. *AIAA J.* **8,** 2282–2284.
Settles, G. S. (1982). Color schlieren optics—A review of techniques and applications. *In* "Flow Visualization II" (W. Merzkirch, ed.), pp. 749–759. Hemisphere, Washington, D.C.
Sivasubramanian, M. S., Cole, R., and Sukanek, P. C. (1984). Optical temperature gradient measurements using speckle photography. *Int. J. Heat Mass Transfer* **27,** 773–780.
Slattery, R. E., Clay, W. G., and Ferdinand, A. P. (1968). High speed photographic techniques in a ballistic range. *Proc. Int. Congr. High-Speed Photogr., 8th* pp. 351–356.
Smith, L. L., and Waddell, J. H. (1970). Techniques of color schlieren. *Proc. Int. Congr. High-Speed Photogr., 9th,* pp. 368–373.
Stanic, S. (1978). Quantitative schlieren visualization. *Appl. Opt.* **17,** 837–842.
Stastný, M., and Pekárek, K. (1985). Flow visualization in a steam turbine profile cascade using a colour schlieren method. *In* "Optical Methods in Dynamics of Fluids and Solids" (M. Pichal, ed.), pp. 265–273. Springer-Verlag, Berlin and New York.
Stevenson, T. N. (1973). The phase configuration of internal waves around a body moving in a density stratified fluid. *J. Fluid Mech.* **60,** 759–767.

Stilp, A. (1968). Der Freiflugkanal des Ernst-Mach-Instituts. *Z. Flugwiss.* **16,** 12–16.
Stolzenburg, W. A. (1965). The double knife-edge technique for improved schlieren sensitivity in low-density hypersonic aerodynamic testing. *J. SMPTE* **74,** 654–659.
Stricker, J., and Kafri, O. (1982). A new method for density gradient measurements in compressible flows. *AIAA J.* **20,** 820–823.
Suchorukich, W. S. (1968). Beugungstheorie der Schlieren. *Proc. Int. Congr. High-Speed Photogr., 8th* pp. 341–345.
Thompson, P. A., Kim, Y.-G., and Meier, G. E. A. (1985). Flow visualization of a shock wave by simple refraction of a back-ground grid. *In* "Optical Methods in Dynamics of Fluids and Solids" (M. Pichal, ed.), pp. 225–231. Springer-Verlag, Berlin and New York.
Thorpe, S. A. (1973). Experiments on instability and turbulence in a stratified shear flow. *J. Fluid Mech.* **61,** 731–751.
Toepler, A. (1864). "Beobachtungen nach einer neuen optischen Methode." Max Cohen & Sohn, Bonn.
Trolinger, J. D. (1974). Laser instrumentation for flow field diagnostics. *AGARDograph* **AGARD-AG-186.**
Uberoi, M. S., and Kovasznay, L. S. G. (1955). Analysis of turbulent density fluctuations by the shadow method. *J. Appl. Phys.* **26,** 19–24.
Vasil'ev, L. A., and Otmennikov, V. N. (1976). Studies of high-velocity gas flows by a shadow photometric method using visualizing diaphragms of complicated shapes. *Sov. J. Opt. Technol.* (*Engl. Transl.*) **43,** 457–459.
Véret, C. (1970). Visualisation à faible masse volumique. *AGARD Conf. Proc.* **AGARD-CP-38,** 257–264.
Weinberg, F. J. (1963). "Optics of Flames." Butterworth, London.
Wernekinck, U., and Merzkirch, W. (1986). Measurement of natural convection by speckle photography. *In* "Heat Transfer 86." (C. L. Tien, V. P. Carey, and J. K. Ferrell, eds.), pp. 531–535. Hemisphere, Washington, D.C.
Wernekinck, U., Merzkirch, W., and Fomin, N. A. (1985). Measurement of light deflection in a turbulent density field. *Exp. Fluids* **3,** 206–208.
Wolter, H. (1956). Schlieren, Phasenkontrast und Lichtschnittverfahren. *In* "Handbuch der Physik" (S. Flügge, ed.), Vol. 24, pp. 555–645. Springer-Verlag, Berlin and New York.

Section 3.3

Achia, B. U., and Thompson, D. W. (1972). Real-time hologram-moiré interferometry for liquid flow visualization. *Appl. Opt.* **11,** 953–954.
Anderson, E. E., Stevenson, W. H., and Viskanta, R. (1975). Estimating the refractive error in optical measurements of transport phenomena. *Appl. Opt.* **14,** 185–188.
Anderson, J. S., Jungowski, W. M., Hiller, W. J., and Meier, G. E. A. (1977). Flow oscillations in a duct with a rectangular cross-section. *J. Fluid Mech.* **79,** 769–784.
Aung, W., and O'Regan, R. (1971). Precise measurement of heat transfer using holographic interferometry. *Rev. Sci. Instrum.* **42,** 1755–1759.
Aung, W., Fletcher, L. S., and Sernas, V. (1972). Developing laminar free convection between vertical flat plates with asymmetric heating. *Int. J. Heat Mass Transfer* **15,** 2293–2308.
Bathelt, A. G., and Viskanta, R. (1980). Heat transfer at the solid–liquid interface during melting from a horizontal cylinder. *Int. J. Heat Mass Transfer* **23,** 1493–1503.
Belotserkovsky, S. M. (1968). Anwendungsmöglichkeiten und Perspektiven optischer

Methoden in der Gasdynamik. *Proc. Int. Congr. High-Speed Photogr., 8th,* pp. 410–414.

Ben-Dor, G., Whitten, B. T., and Glass, I. I. (1979). Evaluation of perfect and imperfect-gas interferograms by computer. *Int. J. Heat Fluid Flow* **1,** 77–91.

Bennett, F. D., Carter, W. C., and Bergdolt, V. E. (1952). Interferometric analysis of air flow about projectiles in free flight. *J. Appl. Phys.* **23,** 453–469.

Black, W. Z., and Carr, W. W. (1971). Application of differential interferometer to the measurement of heat transfer coefficients. *Rev. Sci. Instrum.* **42,** 337–340.

Black, W. Z., and Norris, J. K. (1974). Interferometric measurement of fully turbulent free convective heat transfer coefficients. *Rev. Sci. Instrum.* **45,** 216–218.

Boxman, R. L., and Sloan, M. L. (1978). Scanning technique for obtaining linear fringe shift readout from a high resolution interferometer. *Appl. Opt.* **17,** 2794–2797.

Boyd, R. D., Miller, D. J., and Ghiglia, D. C. (1985). Automated data reduction for optical interferometric data. *In* "Flow Visualization III" (W.-J. Yang, ed.), pp. 140–144. Hemisphere, Washington, D.C.

Brand, B., and Grigull, U. (1974). Interferometrische Beobachtung thermischer Grenzschichten in Gasen bei grösseren Temperaturdifferenzen. *Waerme- Stoffuebertrag.* **7,** 182–188.

Branston, D. W., and Mentel, J. (1976). Beugungstheoretische Behandlung eines Differentialinterferometers für ausgedehnte Phasenobjekte. *Appl. Phys.* **11,** 241–246.

Bryanston-Cross, P. J., Lang, T., Oldfield, M. L. G., and Norton, R. J. G. (1981). Interferometric measurements in a turbine cascade using image-plane holography. *J. Eng. Power* **103,** 124–130.

Bryngdahl, O., and Lee, W.-H. (1974). Shearing interferometry in polar coordinates. *J. Opt. Soc. Am.* **64,** 1606–1615.

Bühler, K., and Oertel, H. (1982). Thermal cellular convection in rotating rectangular boxes. *J. Fluid Mech.* **114,** 261–282.

Burner, A. W., and Goad, W. K. (1981). Phase control during reconstruction of holographically recorded flow fields using real-time holographic interferometry. *NASA Tech. Memo.* **NASA TM-X-81953.**

Buxmann, J. (1970). Messungen von Luftströmungen mit dem Mach-Zehnder-Interferometer. *Forsch. Ingenieurwes.* No. 4, 111–119.

Carlomagno, G. M. (1985). Schlieren interferometry in the mass diffusion of a two-dimensional jet. *Exp. Fluids* **3,** 137–141.

Cha, S., and Vest, C. M. (1981). Tomographic reconstruction of strongly refracting fields and its application to interferometric measurement of boundary layer. *Appl. Opt.* **20,** 2787–2794.

Chevalerias, R., Latron, Y., and Véret, C. (1957). Methods of flow interferometry applied to the visualization of flows in wind tunnels. *J. Opt. Soc. Am.* **47,** 703–706.

Clark, J. A. (1978). Holographic visualization of acoustic fields. *J. Sound Vib.* **56,** 167–174.

Cline, H. E., Holik, A. S., and Lorensen, W. E. (1982). Computer-aided surface reconstruction of interference contours. *Appl. Opt.* **21,** 4481–4488.

Dändliker, R., Marom, E., and Mottier, F. M. (1976). Two-reference-beam holographic interferometry. *J. Opt. Soc. Am.* **66,** 23–30.

Debler, W. R., and Vest, C. M. (1977). Observations of a stratified flow by means of holographic interferometry. *Proc. R. Soc. London, Ser. A* **358,** 1–16.

Decker, A. J. (1981). Holographic flow visualization of time-varying shock waves. *Appl. Opt.* **20,** 3120–3127.

Délery, J., Surget, J., and Lacharme, J.-P. (1977). Interférométrie holographique quantitative en écoulement transsonique bidimensional. *Rech. Aerosp.* **1977-2,** 89–101.

Dullforce, T. A., and Faw, R. E. (1979). High-speed cine recording of real-time holographic interference fringes. *Opt. Commun.* **31,** 111–113.

Dunagan, S. E., Brown, J. L., and Miles, J. B. (1985). A holographic interferometric study of an axisymmetric shockwave/boundary-layer strong interaction flow. *AIAA Pap.* **85-1564.**

Farrell, P. V., Springer, G. S., and Vest, C. M. (1982). Heterodyne holographic interferometry: concentration and temperature measurements in gas mixtures. *Appl. Opt.* **21,** 1624–1627.

Faw, R. E., and Dullforce, T. A. (1981). Holographic interferometry measurement of convective heat transport beneath a heated horizontal plate in air. *Int. J. Heat Mass Transfer* **24,** 859–869.

Flack, R. D. (1978a). Mach–Zehnder interferometry errors resulting from test section misalignment. *Appl. Opt.* **17,** 985–987.

Flack, R. D. (1978b). Shearing interferometer inaccuracies due to a misaligned test section. *Appl. Opt.* **17,** 2873–2875.

Françon, M. (1952). Interférométrie par double réfraction en lumière blanche. *Rev. Opt.* No. 31, 65–80.

Frohn, A. (1967). Measurements of concentration profiles in jets using Fizeau fringes. *AIAA J.* **5,** 185–186.

Gille, J. (1967). Interferometric measurement of temperature gradient reversal in a layer of convecting air. *J. Fluid Mech.* **30,** 371–384.

Gontier, G. (1957). Contribution à l'étude de l'interféromètre différentiel à biprisme de Wollaston. *Publ. Sci. Tech. Minist. Air* (*Fr.*) No. 338.

Gorenflo, R. (1964). Numerische Methoden zur Lösung einer Abelschen Differentialgleichung. Inst. Plasmaphys. Garching, Rep. No. IPP/6/19.

Grigull, U. (1963). Einige optische Eigenschaften thermischer Grenzschichten. *Int. J. Heat Mass Transfer* **6,** 669–679.

Grönig, H. (1967). New MZI-technique for shock-tube measurements. *AIAA J.* **5,** 1046–1047.

Hannah, B. W., and Havener, A. G. (1975). Applications of automated holographic interferometry. *ICIASF '75 Rec.* (*Int. Congr. Instrum. Aerosp. Simul. Facilities*), *IEEE Publ.* **75 CHO 993-6 AES,** 237–246.

Harvey, R. J. (1970). High resolution two-dimensional interferometry. *Rev. Sci. Instrum.* **41,** 1142–1146.

Hatfield, D. W., and Edwards, D. K. (1981). Edge and aspect ratio effects on natural convection from the horizontal heated plate facing downwards. *Int. J. Heat Mass Transfer* **24,** 1019–1024.

Heflinger, L. O., Wuerker, R. F., and Brooks, R. E. (1966). Holographic interferometry. *J. Appl. Phys.* **37,** 642–649.

Herman, G. T. (1980). "Image Reconstruction from Projections." Academic Press, New York.

Howes, W. L. (1984). Large-aperture interferometer with local reference beam. *Appl. Opt.* **23,** 1467–1473.

Howes, W. L., and Buchele, D. R. (1966). Optical interferometry of inhomogeneous gases. *J. Opt. Soc. Am.* **56,** 1517–1528.

Huber, M. C. E. (1971). Interferometric gas diagnostics by the hook method. *In* "Modern Optical Methods in Gas Dynamic Research" (D. S. Dosanjh, ed.), pp. 85–112. Plenum, New York.

Hugenschmidt, M., and Vollrath, K. (1970). Interferometry of rapidly varying phase objects using the fundamental and the harmonic wavelengths of a ruby laser. *Proc. Int. Congr. High-Speed Photogr., 9th,* pp. 86–92.

Hugenschmidt, M., and Vollrath, K. (1981). Light sources and recording methods. *In* "Fluid Dynamics" (R. J. Emrich, ed.), Methods in Experimental Physics, Vol. 18B, pp. 687–753. Academic Press, New York.
Hunter, A. M., and Schreiber, P. W. (1975). Mach–Zehnder interferometer data reduction method for refractively inhomogeneous test objects. *Appl. Opt.* **14,** 634–639.
Jagota, R. C., and Collins, D. J. (1972). Finite fringe holographic interferometry applied to a right circular cone at angle of attack. *J. Appl. Mech.* **39,** 897–903.
Junginger, H.-G., and van Haeringen, W. (1972). Calculation of three-dimensional refractive index field using phase integrals. *Opt. Commun.* **5,** 1–4.
Kahl, G. D., and Mylin, D. C. (1965). Refractive deviation errors of interferograms. *J. Opt. Soc. Am.* **55,** 364–372.
Kaiser, E. (1984). Interferometrische Temperaturfeldmessung für einen thermischen Fluidgeschwindigkeitsaufnehmer. *MSR* (*Berlin*) **27,** 151–154.
Kean, L. (1961). Coefficients for axisymmetric schlieren evaluations. Wright-Patterson AFB, ASD Tech. Note 61-56.
Kelley, J. G., and Hargreaves, R. A. (1970). A rugged inexpensive shearing interferometer. *Appl. Opt.* **9,** 948–952.
Kim, C.-J. (1982). Polynomial fit of interferograms. *Appl. Opt.* **21,** 4521–4525.
Kinder, W. (1946). Theorie des Mach–Zehnder-Interferometers und Beschreibung eines Gerätes mit Einspiegeleinstellung. *Optik* **1,** 413–448.
Kogelschatz, U. (1974). Application of a simple differential interferometer to high current arc discharges. *Appl. Opt.* **13,** 1749–1752.
Kosakoski, R. A., and Collins, D. J. (1974). Application of holographic interferometry to density field determination in transonic corner flow. *AIAA J.* **12,** 767–770.
Koster, J. N. (1983). Interferometric investigation of convection in plexiglas boxes. *Exp. Fluids* **1,** 121–128.
Kraushaar, R. (1950). A diffraction grating interferometer. *J. Opt. Soc. Am.* **40,** 480–481.
Ladenburg, R., and Bershader, D. (1954). Interferometry. *In* "Physical Measurements in Gas Dynamics and Combustion" (R. W. Ladenburg, ed.), pp. 47–78. Princeton Univ. Press, Princeton, New Jersey.
Lee, G., Buell, D. A., Licursi, J. P., and Craig, J. E. (1984). Laser holographic interferometry for an unsteady airfoil undergoing dynamic stall. *AIAA J.* **22,** 504–511.
Mach, E., and von Weltrubsky, (1878). Über die Formen der Funkenwellen. *Sitzungsber. Kais. Akad. Wiss. Wien, Math.-Naturwiss. Kl.* **78,** 551–560.
Mach, L. (1892). Über einen Interferenz-Refraktor, *Z. Instrumentenkd.* **12,** 89–93.
Maddox, A. R., and Binder, R. C. (1969). The use of an improved diffraction grating interferometer. *Appl. Opt.* **8,** 2191–2198.
Maruyama, Y., Iwata, K., and Nagata, R. (1977). Effect of refractive index dicontinuity on the reconstruction of the refractive index field. *Appl. Opt.* **16,** 2034–2035.
Masliyah, J. H., and Nguyen, T. T. (1979). Mass transfer due to an impinging slot jet. *Int. J. Heat Mass Transfer* **22,** 237–244.
Mastin, G. A., and Ghiglia, D. C. (1985). Digital extraction of interference fringe contours. *Appl. Opt.* **24,** 1727–1728.
Matulka, R. D., and Collins, D. J. (1971). Determination of three-dimensional density fields from holographic interferograms. *J. Appl. Phys.* **42,** 1109–1119.
Mayinger, F., and Steinberner, U. (1979). Flow visualization with holographic interferometry. *In* "Flow Visualization" (T. Asanuma, ed.), pp. 341–349. Hemisphere, Washington, D.C.
McLarnon, F. R., Muller, R. H., and Tobias, C. W. (1975). Derivation of one-dimensional refractive-index profiles from interferograms. *J. Opt. Soc. Am.* **65,** 1011–1018.

Mehta, J. M., and Worek, W. M. (1984). Analysis of refraction errors for interferometric measurements in multicomponent systems. *Appl. Opt.* **23,** 928–933.
Merzkirch, W. (1965). A simple schlieren interferometer system. *AIAA J.* **3,** 1974–1976.
Merzkirch, W. (1974). Generalized analysis of shearing interferometers as applied for gas dynamic studies. *Appl. Opt.* **13,** 409–413.
Merzkirch, W., and Erdmann, W. (1973). Evaluation of axisymmetric flow patterns with a shearing interferometer. *Appl. Phys.* **2,** 119–122.
Miller, R. M., and Gebhart, B. (1978). An experimental study of the natural convection flow over a heated ridge in air. *Int. J. Heat Mass Transfer* **21,** 1229–1239.
Miyashiro, S., and Grönig, H. (1985). Low-jitter reliable nanosecond spark source for optical short-duration measurements. *Exp. Fluids* **3,** 71–75.
Montgomery, G. P., and Reuss, D. L. (1982). Effects of refraction on axisymmetric flame temperatures measured by holographic interferometry. *Appl. Opt.* **21,** 1373–1380.
Mowbray, D. E. (1967). The use of schlieren and shadowgraph techniques in the study of flow patterns in density stratified liquids. *J. Fluid Mech.* **27,** 595–608.
Nomarski, G. (1956). Remarques sur le fonctionnement des dispositifs interférentiels à polarisation. *J. Phys. Radium* **17,** 15–35.
Oertel, H. (1967). Messungen im Hyperschallstossrohr. *In* "Kurzzeitphysik" (K. Vollrath and G. Thomer, eds.), pp. 758–848. Springer-Verlag, Berlin and New York.
Oertel, H. (1982). Visualization of thermal convection. *In* "Flow Visualization II" (W. Merzkirch, ed.), pp. 71–76. Hemisphere, Washington, D.C.
Oppenheim, A. K., Urtiew, P. A., and Weinberg, F. J. (1966). On the use of laser light sources in schlieren-interferometer systems. *Proc. R. Soc. London, Ser. A* **291,** 279–290.
Oudin, L., and Jeanmaire, M. (1970). La transformation d'Abel. Applications à la mésure des masses volumiques dans les sillages. Ger.–Fr. Res. Inst. St. Louis, Rep. No. ISL 21/70.
Pandya, T. P., and Weinberg, F. J. (1964). The structure of flat, counter-flow diffusion flames. *Proc. R. Soc. London, Ser. A* **279,** 544–561.
Parks, R. E., and Sumner, R. E. (1978). Bright inexpensive pinhole source. *Appl. Opt.* **17,** 2469.
Pera, L., and Gebhart, B. (1975). Laminar plume interactions. *J. Fluid Mech.* **68,** 259–271.
Philbert, M. (1958). Emploi de la strioscopie interférentielle en aérodynamique. *Rech. Aerosp.* No. 65, 19–27.
Philbert, M., and Surget, J. (1968). Application de l'interférométrie holographique en soufflerie. *Rech. Aerosp.* No. 122, 55–60.
Popovich, M. M., and Weinberg, F. J. (1983). Laser optical methods for the study of very large phase objects. *Exp. Fluids* **1,** 169–178.
Preonas, D. D., and Swift, H. F. (1970). A high-intensity point light source. *Proc. Int. Congr. High-Speed Photogr., 9th,* pp. 148–152.
Reinheimer, C. J., Wiswall, C. E., Schmiege, R. A., Harris, R. J., and Dueker, J. E. (1970). Holographic subsonic flow visualization. *Appl. Opt.* **9,** 2059–2065.
Robinson, D. W. (1983). Automatic fringe analysis with a computer image-processing system. *Appl. Opt.* **22,** 2169–2176.
Rowley, P. D. (1969). Quantitative interpretation of three-dimensional weakly refractive phase objects using holographic interferometry. *J. Opt. Soc. Am.* **59,** 1496–1498.
Sandhu, S. S., and Weinberg, F. J. (1972). A laser interferometer for combustion, aerodynamics and heat transfer studies. *J. Phys. E* **5,** 1018–1020.
Santoro, R. J., Semerjian, H. G., Emmerman, P. J., and Goulard, R. (1981). Optical tomography for flow field diagnostics. *Int. J. Heat Mass Transfer* **24,** 1139–1150.

Schardin, H. (1933). Theorie und Anwendungen des Mach–Zehnderschen Interferenz-Refraktometers. *Z. Instrumentenkd.* **53,** 396–403.
Schardin, H. (1942). Die Schlierenverfahren und ihre Anwendungen. *Ergeb. Exakten Naturwiss.* **20,** 303–439.
Schmidt, F. W., Kulakowski, B., and Wang, D. F. (1984). Evaluation of the effect of variable refraction index on the path of a laser beam. *Exp. Fluids* **2,** 153–158.
Schultz-Grunow, F., and Wortberg, G. (1961). Interferometrische Messungen an einer ebenen laminaren Flamme. *Int. J. Heat Mass Transfer* **2,** 56–80.
Schwarz, G., and Knauss, H. (1982). Quantitative experimental investigation of three-dimensional flow fields around bodies of arbitrary shapes in supersonic flow with optical methods. *In* "Flow Visualization II" (W. Merzkirch, ed.), pp. 737–741. Hemisphere, Washington, D.C.
Sernas, V., and Fletcher, L. S. (1970). A schlieren interferometer method for heat transfer studies. *J. Heat Transfer* **92,** 202–204.
Small, R. D., Sernas, V. A., and Page, R. H. (1972). Single beam schlieren interferometer using a Wollaston prism. *Appl. Opt.* **11,** 858–862.
Smeets, G. (1968). Aufnahmen mit dem Differential-Interferometer und ihre Auswertung. *Proc. Int. Congr. High-Speed Photogr., 8th,* pp. 374–378.
Smeets, G. (1970). Laser-Interferometer zur Messung an schnell-veränderlichen schwachen Phasenobjekten. *Opt. Commun.* **2,** 29–32.
Smigielski, P., and Royer, H. (1968). Application de l'holographie à l'aérodynamique hypersonique en tunnel de tir. *Proc. Int. Congr. High-Speed Photogr., 8th,* pp. 324–327.
Snyder, R., and Hesselink, L. (1984). Optical tomography for flow visualization of the density field around a revolving helicopter rotor blade. *Appl. Opt.* **23,** 3650–3656.
Solignac, J.-L. (1965). Méthode de dépouillement des interférogrammes en écoulement de révolution. *Rech. Aerosp.* No. 104, 12–13.
Solignac, J.-L. (1968). Étude interférométrique de l'écoulement à Mach 5 autour d'une sphère. *Rech. Aerosp.* No. 125, 31–39.
South, R. (1970). An extension to existing methods of determining refractive indices from axisymmetric interferograms. *AIAA J.* **8,** 2057–2059.
Sterret, J. R., Emery, J. C., and Barber, J. B. (1965). A laser grating interferometer. *AIAA J.* **3,** 963–964.
Stevenson, T. N., Woodhead, T. J., and Kanellopulos, D. (1983). Viscous effects in some internal waves. *Appl. Sci. Res.* **40,** 185–197.
Stirnberg, D., Ronkholz, E., and Merzkirch, W. (1983). Der vertikale, laminare, isotherme Freistrahl mit Auftrieb und Diffusion. *Z. Flugwiss. Weltraumforsch.* **7,** 310–315.
Stricker, W. (1984). Analysis of 3-D phase objects by moiré deflectometry. *Appl. Opt.* **23,** 3657–3659.
Surget, J. (1974). Schéma d'holographie à deux sources de référence. *Nouv. Rev. Opt.* **5,** 201–217.
Surget, J., and Chatriot, J. (1969). Cinématographie ultrarapide d'interférogrammes holographiques. *Rech. Aerosp.* No. 132, 51–55.
Sweeney, D. W., and Vest, C. M. (1973). Reconstruction of three-dimensional refractive index fields from multidirectional interferometric data. *Appl. Opt.* **12,** 2649–2664.
Takayama, K., and Onodera, O. (1985). Holographic interferometric study on propagating and focusing of underwater shock waves by micro-explosions. *In* "Optical Methods in Dynamics of Fluids and Solids" (M. Pichal, ed.), pp. 209–216. Springer-Verlag, Berlin and New York.
Tanner, L. H. (1966a). Some applications of holography in fluid mechanics. *J. Sci. Instrum.* **43,** 81–83.

Tanner, L. H. (1966b). The design of laser interferometers for use in fluid mechanics. *J. Sci. Instrum.* **43,** 878–886.
Tanner, L. H. (1968). A study of fringe clarity in laser interferometry and holography. *J. Sci. Instrum.* **1,** 517–522.
Trolinger, J. D. (1979). Application of generalized phase control during reconstruction of flow visualization holography. *Appl. Opt.* **18,** 766–774.
Trolinger, J. D., and Simpson, G. D. (1979). Diagnostics of turbulence by holography. *Opt. Eng.* **18,** 161–166.
van de Weijer, P., and Cremers, R. M. M. (1983). Hook method: improvement and simplification of the experimental setup. *Appl. Opt.* **22,** 3500–3502.
Vest, C. M. (1974). Formation of fringes from projections: Radon and Abel transforms. *J. Opt. Soc. Am.* **64,** 1215–1218.
Vest, C. M. (1975). Interferometry of strongly refracting axisymmetric phase objects. *Appl. Opt.* **14,** 1601–1606.
Vest, C. M. (1979). "Holographic Interferometry." Wiley, New York.
Vest, C. M., and Prikryl, I. (1984). Tomography by iterative convolution: empirical study and application to interferometry. *Appl. Opt.* **23,** 2433–2440.
Vest, C. M., and Sweeney, D. W. (1970). Holographic interferometry of transparent objects with illumination derived from phase gratings. *Appl. Opt.* **9,** 2321–2325.
Vollrath, K., Thomer, G., eds. (1967). "Kurzzeitphysik." Springer-Verlag, Berlin and New York.
Walklate, P. J. (1981). A two wavelength holographic technique for the study of two-dimensional thermal boundary layers. *Int. J. Heat Mass Transfer* **24,** 1051–1057.
Weinberg, F. J. (1963). "Optics of Flames." Butterworth, London.
Weyl, F. J. (1954). Analysis of optical methods. *In* "Physical Measurements in Gas Dynamics and Combustion" (R. W. Ladenburg, ed.), pp. 3–25. Princeton Univ. Press, Princeton, New Jersey.
Winckler, J. (1948). The Mach interferometer applied to studying an axially symmetric supersonic air jet. *Rev. Sci. Instrum.* **19,** 307–322.
Witte, A. B., and Mantrom, D. D. (1975). Interferometric technique for measuring mixing of a buoyant plume. *AIAA J.* **13,** 535–536.
Witte, A. B., and Wuerker, R. F. (1970). Laser holographic interferometry study of high-speed flow fields. *AIAA J.* **8,** 581–583.
Xia, S.-J. (1985). Double mirror laser interferometer. *In* "Flow Visualization III" (W.-J. Yang, ed.), pp. 155–159. Hemisphere, Washington, D.C.
Yokozeki, S., and Mihara, S. (1979). Moiré interferometry. *Appl. Opt.* **18,** 1275–1280.
Zehnder, L. (1891). Ein neuer Interferenzrefraktor. *Z. Instrumentenkd.* **11,** 275–285.
Zien, T. F., Ragsdale, W. C., and Spring, W. C. (1974). Quantitative determination of three-dimensional density field of a circular cone by holographic interferometry. *AIAA Pap.* **74-636.**
Zien, T. F., Ragsdale, W. C., and Spring, W. C. (1975). Quantitative determination of three-dimensional density field by holographic interferometry. *AIAA J.* **13,** 841–842.

Section 3.4

Anderson, R. C., and Taylor, M. W. (1982). Phase contrast flow visualization. *Appl. Opt.* **21,** 528–536.
Erdmann, S. F. (1951). Ein neues, sehr einfaches Interferometer zum Erhalt quantitativ auswertbarer Strömungsbilder. *Appl. Sci. Res., Sect. B* **2,** 1–50.

Peterka, J. A., and Richardson, P. D. (1984). Effects of sound on local transport from a heated cylinder. *Int. J. Heat Mass Transfer* **27,** 1511–1523.
Philbert, M. (1964). Visualisation des écoulement à basse pression. *Rech. Aerosp.* No. 99, 39–48.
Véret, C. (1970). Visualisation à faible masse volumique. *AGARD Conf. Proc.* **AGARD-CP-38,** 257–264.
Wolter, H. (1956). Schlieren, Phasenkontrast und Lichtschnittverfahren. *In* "Handbuch der Physik" (S. Flügge, ed.), Vol. 24, pp. 555–645. Springer-Verlag, Berlin and New York.

Section 3.5

Boeder, P. (1932). Über Strömungsdoppelbrechung. *Z. Phys.* **75,** 258–278.
Boyer, G. R., Lamouroux, B., and Prade, B. S. (1978). Atmospheric birefringence under wind speed gradient shear. *J. Opt. Soc. Am.* **68,** 471–474.
Durreli, A. J., and Norgard, J. S. (1972). Experimental analysis of slow viscous flow using photoviscosity and bubbles. *Exp. Mech.* **12,** 169–177.
Horsmann, M., and Merzkirch, W. (1981). Scattered light streaming birefringence in colloidal solutions. *Rheol. Acta* **20,** 501–510.
Horsmann, M., Schmitz, E., and Merzkirch, W. (1979). The application of streaming birefringence to the quantitative study of low-Reynolds number pipe flow. *In* "Flow Visualization" (T. Asanuma, ed.), pp. 369–375. Hemisphere, Washington, D.C.
Kuske, A., and Robertson, G. (1974). "Photoelastic Stress Analysis." Wiley, London.
Liepsch, D., Moravec, S., and Zimmer, R. (1982). Visualization of stationary and pulsating flow in artery models. *In* "Flow Visualization II" (W. Merzkirch, ed.), pp. 587–591. Hemisphere, Washington, D.C.
Mach, E. (1873). "Optisch-akustische Versuche (Die spectrale und stroboskopische Untersuchung tönender Körper)." Calve, Prague.
Maxwell, J. C. (1873). Double refraction of viscous fluids in motion. *Proc. R. Soc. London, Ser. A* **22,** 46–47.
McAfee, W. J., and Pih, H. (1971). A scattered light polariscope for three-dimensional birefringent flow studies. *Rev. Sci. Instrum.* **42,** 221–223.
McAfee, W. J., and Pih, H. (1974). Scattered-light flow-optic relations adaptable to three-dimensional flow birefringence. *Exp. Mech.* **14,** 385–391.
Nakatani, N., Yamada, T., and Soezima, Y. (1971). Development of a new aqueous solution highly sensitive to flow birefringence. *Jpn. J. Appl. Phys.* **10,** 1034–1039.
Peebles, F. N., and Liu, K. C. (1965). Photoviscous analysis of two-dimensional laminar flow in an expanding jet. *Exp. Mech.* **5,** 299–304.
Philippoff, W. (1964). Streaming birefringence of polymer solutions. *J. Polym. Sci., Part C* **5,** 1–9.
Pindera, J. T., and Krishnamurthy, A. R. (1978a). Characteristic relations of flow birefringence. Part 1: Relations in transmitted radiation. *Exp. Mech.* **18,** 1–10.
Pindera, J. T., and Krishnamurthy, A. R. (1978b). Characteristic relations of flow birefringence. Part 2: Relations in scattered radiation. *Exp. Mech.* **18,** 41–48.
Prados, J. W., and Peebles, F. N. (1959). Two-dimensional laminar-flow analysis, utilizing a double refracting liquid. *AIChE J.* **5,** 225–234.
Schmitz, E., and Merzkirch, W. (1983). Direct interferometric measurement of streaming birefringence. *Rheol. Acta* **22,** 75–80.

Schmitz, E., and Merzkirch, W. (1984). A test fluid for simulating blood flows. *Exp. Fluids* **2,** 103–104.

Swanson, M. M., and Green, R. L. (1969). Colloidal suspension properties of milling yellow dye. *J. Colloid Interface Sci.* **29,** 161–163.

Wayland, H. (1964). Streaming birefringence as a rheological research tool. *J. Polym. Sci., Part C* **5,** 11–36.

4

Flow Field Marking by Heat and Energy Addition

4.1. Artificially Introduced Density Changes

Chapters 2 and 3 deal with two basic, different principles for visualizing flows: The first is to introduce into the fluid a foreign substance being visible and moving with the flow, and from the motion of this substance draw conclusions as to the motion of the fluid; the second principle applies to the motion of fluids having variations in their density; these variations may be visualized by optical means owing to a relationship between the fluid density and the refractive index of the fluid. The different ranges of applicability of these two visualization principles coincide roughly with the classes of incompressible and compressible flows. In the following sections we will discuss a third group of visualization techniques, which can be regarded to be a combination of the two former principles. The foreign substance, in this case, is immaterial, it is energy, which will transferred to certain portions of the flow. These portions or fluid elements behave like tracers in the flow and can be discriminated from the rest of the fluid owing to their increased energy level. The way of discrimination depends on the type of flow and on the rate of energy introduced into the fluid.

By adding the energy at singular points to an incompressible flowfield, the flow is artificially made to have density variations. Such portions of the fluid having an altered density level can either be visualized by one of the established optical methods or can be directly observed due to a certain degree of luminosity of the respective fluid element, if the rate of energy transferred to a flow is high. Such a luminosity is associated not only with the generation of strong sparks in a gas flow, but also with the electron-beam and the glow-discharge methods. These two latter methods are mostly used for visualization of flows, which are compressible, and which also have a very low average density level, so that the absolute density changes are too small to be identified with an optical method. Hence, with this kind of energy addition one has found a way to visualize a third class of flows, the flow or rarefied gases, which is for several reasons distinguished from the "ordinary" compressible flows. It should finally be mentioned that a gas flow having an extremely high level of kinetic energy becomes luminous, if the flow is brought to rest in a stagnation regime where the kinetic energy is transferred into heat. The released heat must generate gas temperatures high enough to excite electronic transitions in the gas. Such a flow is visible *per se* (Fig. 4.1) and does not need to be visualized.

Fig. 4.1 Shock-tunnel air flow of hypersonic Mach number around a blunt test model. Temperatures in the stagnation zone in front of the model exceed 4000 K, so that the gas in this flow regime becomes luminous. (Courtesy of NASA Ames Research Center.)

A simple way of adding heat to an incompressible gaseous flow is to brace an electrically heated thin wire across the stream. The temperature of the fluid elements passing close to the wire is raised, and, since the pressure in the flow remains constant, their density is lower than that of the surrounding, undisturbed flow. The density difference can be controlled with the electric current. Schardin (1942) demonstrated that, with a grid of such wires, one may observe filament lines in a two-dimensional, low-speed air stream. The grid is in a plane perpendicular to the flow direction, and the filament lines are visualized with the aid of a schlieren system having the light beam parallel to the wires of the grid. As defined earlier, filament lines consist of fluid particles having all passed the same fixed point in the flow field, in this case the heated wire. In order to discriminate the filament lines over a greater distance downstream of the wire, it is required that the rate of diffusion of the heated gas into the undisturbed flow be low, that is, that the ratio of diffusive velocity to flow velocity remain small. This condition can only be fulfilled in a laminar flow; the method is therefore appropriate to detect in a stream laminar-to-turbulent transitions, which would appear on a schlieren picture in form of a sudden decay of the filament lines. If one uses a periodically pulsed current instead of a stationary current for heating the wire, one may produce "hot spots" in the flow, which are also observed with the aid of a schlieren system (Dewey, 1973). A schlieren photograph made with a sufficiently short exposure time shows several hot spots in the flow field, which have been generated by the repetitive heating of the wire. Since the time interval between electric pulses is known, one measures the flow velocity from the distance of the hot spots on the photographs. In Section 4.2 we will discuss the fact that such velocity measurements may contain severe errors, if the flow is accelerated. In order to generate sharp hot spots it is necessary that the wire cools down rapidly after each heating pulse (Fig. 4.2).

More frequently than for measuring velocity fields has the generation of weak refractive index variations been used for visualizing vortex sheets and mixing layers. Two typical experimental procedures will be mentioned here. Ohashi and Ishikawa (1972) study the vortex shedding from an oscillating airfoil in uniform, low-speed airflow. The upper surface of the airfoil is painted black and directly heated by the radiation of an infrared lamp. The air stream passing over this surface is heated, and a strong temperature gradient or density gradient develops in the dividing stream line between the two streams separating from the lower and the upper surface. This line, which is regarded the center of the vortex sheet, is made visible in a schlieren system. Pierce (1961) investigates the shedding and the rolling up of a vortex sheet that separates from a flat plate

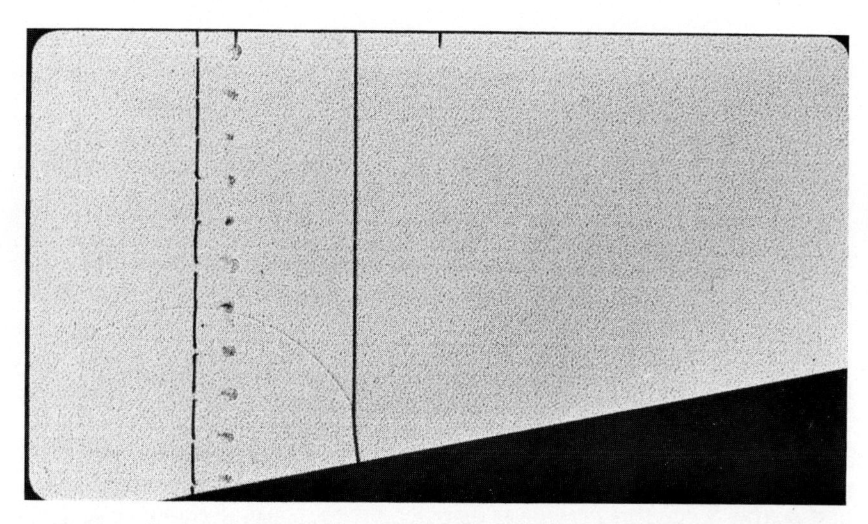

Fig. 4.2 Shadowgraph of "hot spots" generated in a two-dimensional shock tube flow. The heating wire is perpendicular to the flow direction. Hot spots are generated from the nonisolated portions of the wire by a single electric pulse. Time interval between electric pulse and photographic exposure is determined from the position of the shock front and the known shock speed. (Courtesy of J. M. Dewey, University of Victoria, British Columbia.)

accelerated normal to itself from rest in still air. The wooden plate is soaked with benzene, which evaporates into the air stream passing close to the plate surface. Benzene vapor fills the vortex, which is visualized in a shadowgraph due to the refractive index differences between air and the vapor.

From Section 2.4.3 we know that temperature differences can be picked up and transformed into a visible pattern by means of an IR camera. An interesting application of this instrument to a flow situation with temperature differences introduced by purpose has been reported by Brydon *et al*. (1979). The flow of a cold saline through the vessel in the warm environment of the myocard is viewed with an IR camera during bypass surgery, and it allows the surgeon for checking the flow conditions after the operation.

4.2. Velocity Mapping with Spark Tracers

The high voltage applied between two electrodes in a gaseous flow may produce an electric spark discharge, which can be used for marking and visualizing certain flow elements. The fluid element ionized and thereby illuminated by the discharge is the visible tracer in the flow. The generation of the discharge in a direction normal to the main flow direction allows for mapping the velocity distribution in certain flow fields.

We consider two electrodes with a spatial distance of several centimeters in the gas flow (Fig. 4.3). The first passage of a spark after a discharge creates in the gas an ionized path or column, which persists for a period of time of the order of 100 μsec, depending on the discharge conditions. This column is swept downstream by the mean flow. Since the plasma column exhibits a smaller electric resistance than the surrounding neutral gas, a second spark produced during the lifetime of the column would rather follow this preionized path traced by the first spark, than take the shortest and most direct way between the electrodes. If a series of sparks of short duration is produced across the flow at intervals smaller than the above-mentioned plasma lifetime of about 100 μsec, each spark of the series will follow and thereby reilluminate the column traced by the first spark. Since this column is displaced with nearly the flow velocity, a profile of this displacement by the flow can be obtained by means of an open-shutter photograph. From the known frequency of the spark series and the measured displacement one may derive the local flow velocity. This is analogous to the generation of time lines with the hydrogen bubble technique (Section 2.3.2) or with the smoke wire (Section 2.1.3.3).

The spark-tracer technique has been developed and adapted to the study of boundary-layer flow by Bomelburg *et al*. (1959). Improvements of the electric impulse generator, which allowed for higher values of the applied voltage and for shorter discharge times, have been reported by Früngel (1960) and Früngel and Thorwart (1970). The technique has been reexamined and reviewed by Früngel (1977), Asanuma *et al*. (1979), and Nakayama *et al*. (1979).

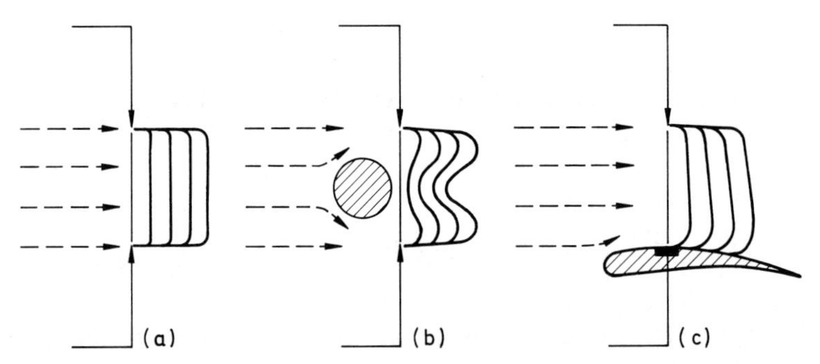

Fig. 4.3 Principle of spark-tracer technique applied to three different flows: (a) uniform, parallel flow, (b) wake flow behind a sphere, and (c) flow over an airfoil.

For the purpose of visualizing velocity profiles, the first spark of the series is desired to trace along a straight line between the electrodes. Since the first spark develops in the direction of the gradient of the electric potential, the condition of a straight spark trace can be met by using parallel plate electrodes. Such electrodes, however, would cause an appreciable interference with the flow. Spiked electrodes, as sketched in Fig. 4.3, generate a minimum of aerodynamic interference, but do not give to the spark a defined initial direction. The design of the electrodes is therefore a compromise between electrical and aerodynamic requirements. With two electrodes having the form of a rod (Fig. 4.4), the discharge of each spark starts from a different point along the rod, according to the momentary position of the preionized gas column. The possible distance between electrodes depends on the pressure of the free flow and the voltage applied; usual distances are up to 10 or 12 cm at discharge tensions of up to 250 kV. This tension is applied to the capacitors, which are discharged across the spark gap. Since usual frequencies for the sparks in the series are about 100 kHz, the capacitors must be recharged within only a few microseconds. The discharge time is below 1 μsec. Such electric systems, which supply the necessary charge within a short enough time interval, and which control the frequency of the spark discharge, are now commercially available.

The appropriate spark frequency depends on the mean velocity of the flow. The smallest possible frequency or maximum distance between sparks at flow velocities is determined by the diffusion of the particles from the ionized gas column into the ambient gas (i.e., by the lifetime of the ionized column). This lifetime might not be significant for determining the spark frequency at high flow velocities: If the interval between sparks is too long and the flow velocity high, the ionized column will be displaced too great a distance to have the next spark follow it, and instead, each successive spark will trace a new path. This may lead to severe errors if flows with a large velocity gradient perpendicular to the mean flow direction (e.g., wake flows) are to be investigated. If the spark frequency is chosen to meet the appropriate conditions for the low-speed portion of the wake flow, this might be too low for the higher wake velocities, and the measured velocity values will be too small in this region. Another error source in this connection is turbulent fluctuations of the flow. The spark intervals are usually smaller than the temporal scale of turbulence; the fluctuations do not average out, and the measured profiles exhibit velocity variations caused by these turbulent fluctuations.

The spark-tracing technique precipitates two questions concerning reliability and precision: How does the energy of the sparks, which is transferred to the surrounding gas, disturb the mean flow properties, and does

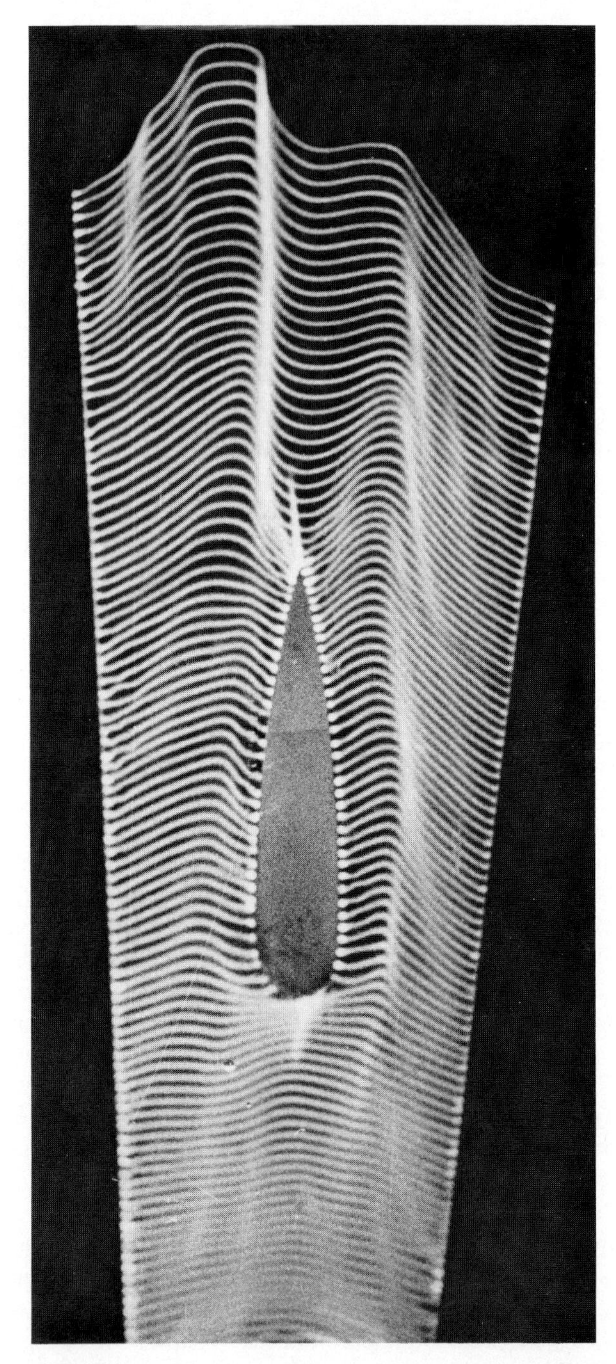

Fig. 4.4 Laminar flow around a metallic profile. The model is placed between two rod electrodes. (From Früngel, 1977.)

the ionized gas column exactly follow the flow? Früngel (1970) investigated the first question, and according to this estimates, the total thermal energy added to the flow is, in most cases of application, only a small fraction of the kinetic energy of the flow. The second problem, which is the same as for any other type of tracer, particle, or bubble, has extensively been studied by Rudinger and Somers (1960). A general rule, as discussed in Section 2.2.1, is that tracers, which have a density different from that of the surrounding fluid, can move exactly with the flow only if the flow velocity is constant. In an accelerated flow the velocities of tracer and surrounding fluid should be different. The density of the ionized gas column is smaller than that of the flowing gas, and as a result, the velocity of this ''gaseous tracer'' may exceed the mean flow velocity, if the flow is accelerating or if it has a velocity gradient in the direction of the mean flow. This behavior may be explained by the difference of the inertial masses of the spark-heated column and an equivalent column of the surrounding gas. Beyond this, the relative motion of the ionized column with respect to the mean flow transforms the column into a vortex. This transformation consumes a large fraction of the energy of the relative motion, and its effect is therefore to reduce the initial velocity discrepancy, i.e., to have some kind of damping effect. The remaining energy, however, is still sufficient to cause, in certain cases, an appreciable residual velocity difference.

It is then obvious that the results of tracer-spark velocity measurements in accelerated flows might be associated with large errors. The value of the velocity lag increases with the magnitude of the acceleration and with the ratio of the density of the surrounding gas to the density of the spark-heated column. Because of the latter, the energy released at each discharge should be kept as small as possible. The greatest acceleration is caused by a shock wave. Rudinger and Somers have found, that under these most unfavorable conditions the column velocity may exceed the gas velocity by as much as 60% immediately after the passage of the shock. This difference is almost zero if the spark discharge is generated only after the passage of the shock. Matsuo *et al*. (1981) investigate, both theoretically and experimentally, the measuring error when the spark-tracer method is applied to the accelerating flow of the expansion fan in the driver section of the shock tube, and to the flow in a subsonic nozzle. Surprisingly, the error in the nozzle flow increases with time, and it can become appreciably high in both cases.

The spark-tracer technique has been applied to flow situations that are inappropriate for being investigated with the smoke wire due to relatively high gas velocities. This comprises the flow field in high-enthalpy or high-Mach number flow facilities (Kyser, 1964; Lahye *et al*., 1967; McIntosh,

Fig. 4.5 Spark tracer photograph of the relative flow between the blades of a rotating impeller. (From Fister *et al.*, 1982. Published by Hemisphere Publishing Corporation.)

1971; Kimura *et al.*, 1977; Matsuo *et al.*, 1979), in flow machinery test models (Nakayama *et al.*, 1979; Fister *et al.*, 1982), and particle-laden flows (Bernotat and Umhauer, 1973). The visualization of the air flow between the blades of a rotating radial impeller (Fig. 4.5) makes it necessary to freeze the image of the rotating flow field onto the stationary film of the camera. Fister (1966) has found an interesting solution to this problem: The image of an object is formed by passing the light rays through a Dove prism (Fig. 4.6). If the object is rotated by 180° and the Dove prism by 90°, the orientation of the image will be the same as before the rotation. From this it follows that the image of the rotating object remains stationary, if the prism is rotated at half the speed of the object.

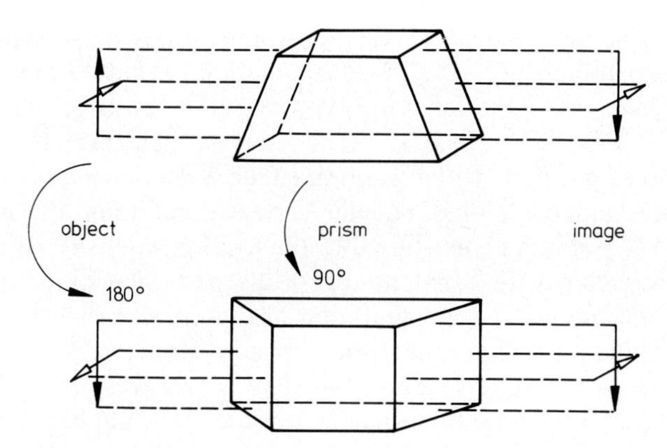

Fig. 4.6 Formation of an image through a Dove prism. The image remains stationary when the prism is rotated by half the angle of rotation of the object. (From Fister *et al.*, 1982. Published by Hemisphere Publishing Corporation.)

4.3. Low-Density Flow Visualization

The optical visualization methods, which make use of the refractive behavior of the gas flow to be studied exhibit a certain sensitivity limit if the average level of the gas density becomes too low. It is in this range of low-density or rarefied gas flows that a visualization of the flow can be achieved by making use of the radiative characteristics of some gases. By means of an appropriate energy release, the molecules of the flowing gas are excited to emit a characteristic radiation. Two different ways of adding the energy to the flow, by an electron beam and by an electric glow discharge, will be discussed here. The intensity of the emitted radiation increases with the value of the local gas density, so that it becomes possible to discriminate between flow regimes of different gas density. For several reasons the range of useful application of these methods is restricted to low gas density levels.

4.3.1. Excitation by Electron Beams

The possibilities of visualizing gas flows at low densities by excitation with an electron beam have been explored first by Schumacher 1953 (see also Grün *et al.*, 1953). A narrow beam of high-energy electrons traverses the gas flow under study; owing to inelastic collisions between the fast electrons and the gas molecules, some gas molecules are excited and subsequently return to the ground state with emission of characteristic radiation. The light emission can be prompt, or it may occur from an excited metastable state. The prompt radiation is emitted more or less at the same place where the gas is excited (i.e., at the position of the electron beam in the flow). The electron beam appears, therefore, as a column of bright fluorescent light, which is often called a *fluorescence probe* (Schumacher and Gadamer, 1958). Under certain conditions, the intensity of the direct radiation is proportional to the local gas density. If one moves the electron beam with constant speed in a particular plane through the gas flow, one obtains a representation of the density distribution in this plane by taking a photographic time exposure while the beam is moving. The lifetime of an excited metastable state is relatively long. The transition into the ground state takes place after the molecule is swept a certain distance by the flow. The associated radiation is emitted at some point in the flow downstream of the original beam position and is called the afterglow radiation. The luminescence of the afterglow radiation is also appropriate for visualizing density changes in the gas flow. It is then not required to move the electron beam, but the intensity of this radiation is much smaller than that of the direct radiation.

Beyond the application for pure flow visualization, the electron beam technique can be used for quantitative temperature and density measurements if it is combined with spectroscopic analysis of the electron beam radiation (Muntz, 1968, 1981). The intensities of a single line or a band in the radiation spectrum are proportional to the number density of the test gas particles, the factor or proportionality depending on both vibrational and rotational temperature of the gas. The measurement of line and band intensities allows one, therefore, to determine vibrational and rotational temperatures, and the concentration rates or partial densities of the active gas species as well. Our discussion of this section, however, is restricted to the sole purpose of producing density sensitive pictures of a rarefied gas flow.

The test gas in most studies is of course air, but only the interaction between fast electrons and nitrogen molecules accounts for the visualization of air flows by means of the fluorescence probe. Most of the N_2 molecules undergoing a collision are ionized and simultaneously excited; the resulting state may be denoted by N_2^{+*}. Provided that the kinetic energy of the electrons is high enough, the most preferred transition is to a level 18.7 eV above the ground level of N_2. The predominant subsequent emission is caused by a spontaneous transition to a level 3.1 eV below the N_2^{+*} state, equivalent to the first negative emission system of N_2^+. The most intensive radiation of this transition is the (0,0) band at a wavelength of 391.4 nm. A first-order analysis shows that the intensity of the radiation emitted per unit length of the electron beam and at constant electric current of the beam is proportional to the number density of the gas molecules in the respective beam section. The proportionality factor is of the type of a collision cross section. Such analysis, however, suffers from several simplifying assumptions. Collision cross sections for all possible transitions which contribute to the total radiation are not known; the measured radiation also contains contributions from collisions of the gas particles with secondary electrons, while the theory cannot account for those electrons, which excite metastable states and do not contribute to the direct radiation. These and other sources (e.g., beam broadening, electron scattering, and quenching collisions) increase at higher gas densities, so that the electron beam technique must be restricted to the investigation of low-density gas flows. Electron beam flow visualization will remain a qualitative method, allowing one just to discriminate between regimes of reduced or increased gas density.

A thin and narrow electron beam of about 1 mm in diameter has to be produced by an appropriate electron gun. Usual values for voltage and current are 20 KV and 1 mA. The test chamber of the wind tunnel and the attached electron gun form one evacuated system. The beam can be moved either mechanically parallel to itself (Rothe, 1965) or by means of deflection coils to cover a certain angular section (Weinstein *et al.*, 1968; Léwy, 1970; see Fig. 4.7). The speed of motion depends on the available test time of the wind tunnel flow. Facilities producing a stationary flow allow a slow movement of the beam, and exposure times up to 60 sec have

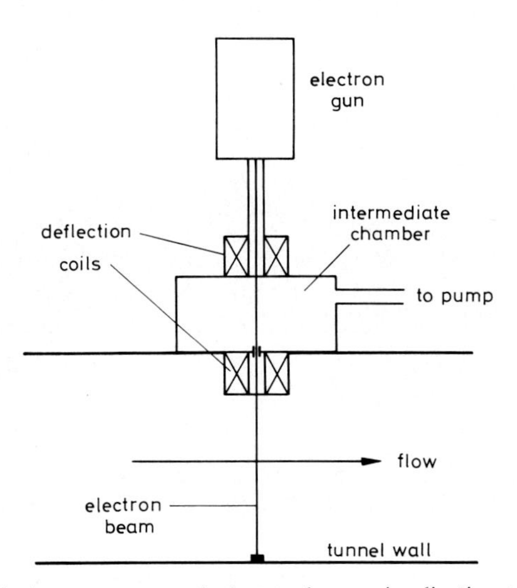

Fig. 4.7 Principal arrangement of electron-beam visualization technique. Electron beam is moved by means of deflection coils.

been used. In order to prevent the production of secondary electrons, the electron beam must be received by a graphite target. Test models in the wind tunnel should be metallic to avoid fluorescence from body surfaces, and the models should be connected to the ground so that no electric charges are built up at these bodies. The direct radiation allows one to visualize supersonic flow fields at a density, which is one or two orders of magnitude below the sensitivity limit of a schlieren system (Fig. 4.8).

The excitation of metastable states in the test gas can cause a noticeable afterglow radiation downstream of the electron beam. This afterglow is less intense than the direct radiation, and only cold flows of nitrogen and argon and mixtures of nitrogen and noble gases yield an afterglow, which is intense enough for taking flow pictures. Flow regimes with an increased gas density can be discriminated due to a more intense afterglow radiation (Sebacher, 1966; Léwy, 1970). The afterglow disappears almost completely in air due to inelastic collisions (quenching) between excited N_2 molecules and nonexcited O_2 molecules. The mechanisms of the transition from the metastable states and the associated emission of radiation is not yet fully understood in this case, and no analysis is available for deriving quantitative data from the flow pictures. An additional difficulty in interpreting the visualized pattern is that the intensity of the

Fig. 4.8 Supersonic air flow around a 20° wedge. Direction of the electron beam is from above to below; a shadow is seen therefore below the test model. The visualization of the shock wave in this shaded regime is due to scattered electrons. (From Léwy, 1970.)

afterglow radiation decreases with increasing distance from the electron beam. The sole advantage of this method is that it is not necessary to move the electron beam through the flow field under study.

4.3.2. Electric Glow Discharge

The electric discharge in gases at low pressures is accompanied by the emission of light. Since the intensity of this radiation depends on the density of the gas in the control volume, one may adapt this method to the visualization of rarefied gas flows. The processes in the glow discharge are similar to those of the electron-beam technique. Free electrons and ions, which are in the test volume, are accelerated by the external electric field and can produce a cascade of secondary electrons and ions due to collisions with neutral gas molecules. The primary and secondary electrons and ions excite gas molecules, which subsequently emit radiation upon spontaneous transition into the ground state. This radiating regime in the electric discharge is called the positive column. The emission intensity of the positive column is a function of the gas density. In a certain density range, the emitted light intensity increases with the number of exciting collisions, and therefore, with the level of the gas density. This, however, holds only up to a particular value of the gas density where the free path length of the electrons becomes too small, and the electrons gain insufficient energy between collisions for excitation. This useful range of radiation is usually at values of about 10^{-3}–10^{-4} of the density at normal conditions.

In order to visualize the compressible flow in a low-density wind tunnel, the test model is made one of the electrodes for the discharge, and a certain portion of the wind tunnel wall may serve as the second electrode (Fig. 4.9). By a suitable choice of the geometry of the electrodes, the field of the positive column can be varied so as to cover the desired portion of the flow field. The potential required between the electrodes depends on the test gas. Appropriate voltages are 1000 V for air or nitrogen and 300 V for helium flows. In order to obtain a uniform luminosity of the positive column, one uses an ac rather than a dc voltage. With the voltage applied between the electrodes, the flowing and radiating gas can be observed or photographed. Density changes in the flow appear as a change in intensity and sometimes in color of the emitted radiation. This method has been applied to visualizing flows of nitrogen (McCroskey *et al.*, 1966), helium (Horstmann and Kussoy, 1968; see Fig. 4.10), and air (San and Ge, 1982).

The discharge spectrum in air is mainly that of nitrogen, but air is not very appropriate for studies with the positive column, since the electric

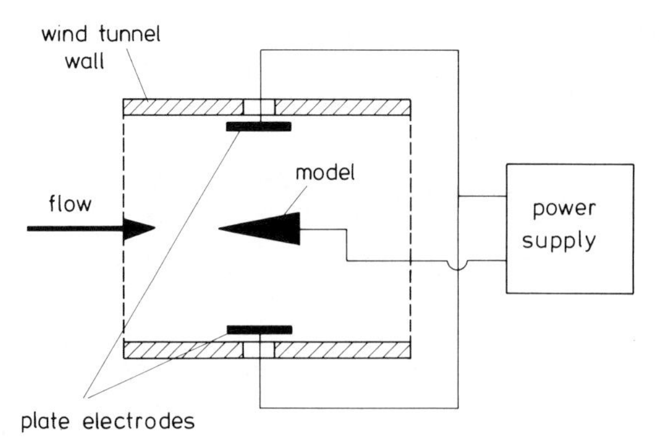

Fig. 4.9 Glow discharge apparatus applied to a low-density supersonic wind tunnel. (See, e.g., San and Ge, 1982).

discharge in air is followed by a great degree of afterglow radiation. The origin of this afterglow can be the excitation of metastable states, as in the case of electron-beam flow visualization. If mixtures of gases are used, the afterglow can also be caused by slow chemical reactions between

Fig. 4.10 Glow discharge photograph obtained in a helium wind tunnel with a free-stream Mach number of 40. The test model is a slender cone with spherical nose. Glow appears light violet on a dark blue background. (Courtesy of E. E. Horstmann, NASA Ames Research Center.)

different chemical constituents of the gases with the associated emission of radiation ("chemiluminescence"). Winkler (1954) has reviewed such possible molecular transitions causing chemiluminescence afterglow in air.

It should finally be mentioned that chemiluminescence can be generated by chemical reactions between certain gaseous components, without external excitation (e.g., the reaction between NO and O or O_3 (ozone) yielding NO_2).

This has been used for visualization of mass transfer and mixing (see, e.g., Hartunian and Spencer, 1966; Van der Bliek *et al.*, 1967; Dickerson and Stedman, 1982), and there is some equivalency to those methods making use of chemically generated color changes in streams of mixing liquids (see Fig. 2.4).

4.4. Laser-Induced Fluorescence

In Sections 2.1.2 and 2.2.2 it was reported that fluorescent dye or tracer particles can be used in liquid flows in order to enhance the visibility in flow visualization studies. These tracers consist of a material whose molecules emit a characteristic, fluorescent radiation upon excitation by light of an appropriate wavelength. The wavelength of fluorescence is different from the wavelength necessary for excitation, and the process of fluorescence, therefore, is governed by inelastic scattering. In this section we will briefly discuss the possibility of generating fluorescence in gas flows, which, in most cases, have been seeded with a suitable tracer material. Excitation of the fluorescence is induced by a laser light source. The abbreviation LIF (laser-induced fluorescence) has been used by a number of authors.

The method is applied to gas flows, mostly supersonic, whose density is below the sensitivity limit of the conventional optical methods described in Section 4.3, or to combustion processes, which develop fluorescent reactants. We will first discuss the case of a gas flow seeded with a gaseous fluorescent tracer. Seeding is performed with molecular iodine I_2 (Rapagnani and Davis, 1979; Cenkner and Driscoll, 1982; Hiller and Hägele, 1982; McDaniel *et al.*, 1982; McDaniel, 1983; Ackermann *et al.*, 1985; Hiller and Hanson, 1985), or with atomic sodium, Na (Miles *et al.*, 1978; Zimmerman *et al.*, 1985). The tracer gas can be excited by a laser beam, which may be directed through the flow field like the electron beam in the experiments reported in the previous section. The tracer molecules thus are pumped to a higher electronic energy level, from which they spontaneously decay to an intermediate energy level, so that the emitted fluorescent radiation is of a wavelength different from that of the exciting light. An appropriate narrow optical filter allows for blocking off all the noise generated by the incident light. From a first-order analysis it follows that the intensity of the emitted fluorescence is proportional to the number density of the tracer molecules in the scattering volume. Since one generally assumes that the tracer molecules are uniformly distributed in the flowing gas, one concludes from the recorded light intensity onto the local value of the gas density in the compressible flow. The fluorescent signal also contains information on the local pressure, temperature, and even on the flow velocity; the latter is due to a Doppler shift of the fluorescent wavelength. Our interest here is mostly in the possibility of visualizing the compressible flow, and for this purpose it is more appropriate to apply the exciting laser light in form of a thin, plane sheet, and not as a single beam ("optical probe"). Fluorescent scattering is then observed in a direction normal to the light sheet, and information is available on the distribution of the gas density in the thin illuminated plane.

Iodine can be mixed easily with the test gas at room temperature. Gasdynamic experiments have been performed with both helium and nitrogen, which do not react with this tracer. Iodine, however, is incompatible with many materials, and it is advisable to use stainless steel, particularly for the seeding apparatus. Seeding rates are of the order of 10^{-3}–10^{-4} per one part of the test gas. The green (514.5 nm) line of an argon ion laser is used for excitation. The emitted fluorescence is yellow. McDaniel (1983) found that most of the dependence of the fluorescent signal on pressure and temperature can be removed by detuning the laser from the absorption frequency by means of an etalon. The signal intensity then is directly proportional to the gas density. A disadvantage is that the mean signal intensity decreases with $1/\Delta\nu^2$, where $\Delta\nu$ is the frequency shift in detuning from the line center. The unwanted influence of pressure and temperature on the fluorescent signal and the dependence of this influence on $\Delta\nu$ imply that the visible pattern is different for different values of the laser wavelength (Fig. 4.11). This change in the pattern is additionally affected by the Doppler shift that is the strongest when the fluorescence is excited near the center of the absorption line of iodine.

Sodium has a number of advantages over iodine as a fluorescent tracer in gas flows. The signal intensity can be up to three orders of magnitude higher at a given laser intensity. Also, the molecular mass of sodium vapor is approximately equal to that of gases to which it has been applied (nitrogen and helium), while the molecular mass of iodine is about one order of magnitude higher. On the other hand, iodine has a higher vapor pressure at room temperature, which simplifies the seeding. For gas dynamic experiments performed with helium, Miles et al. (1978) describe a

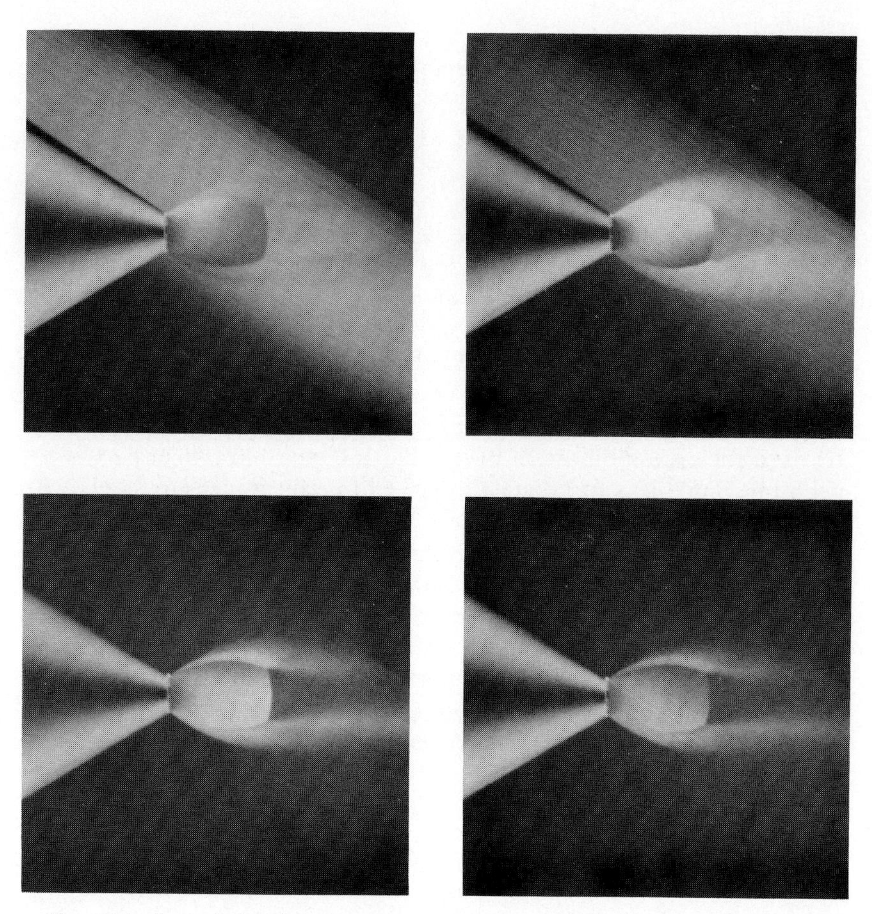

Fig. 4.11 Laser-induced fluorescence patterns of an iodine-seeded underexpanded jet photographed for four different wavelengths of the exciting laser. (From McDaniel, 1983. Copyright © American Institute of Aeronautics and Astronautics; reprinted with permission.)

seeding apparatus in which a hot sodium–helium mixture is produced by purging pressurized helium through an oven where the helium mixes with sodium vapor. The mixture is injected into the helium stagnation chamber upstream of the Laval nozzle, and the relative mass concentration of sodium in the helium flow is estimated to be less than 10^{-3}. Fluorescence is excited by a Rhodamine 6G dye laser, which is tuned through one of the strong absorption frequencies in the red–orange part of the visible spectrum.

The possibility of seeding a flow with a fluorescent tracer, whose radiation intensity is proportional to the number density of the tracer molecules in the scattering volume, is of course not restricted to gaseous flows. Koochesfahani and Dimotakis (1985) have performed concentration measurements in mixing liquid streams, one of which was seeded with sodium fluorescein dye. Fluorescence was excited with an argon ion laser, and the local dye concentration was taken proportional to the visible fluorescence intensity.

When fluorescent reactants are generated in a flame, laser induced fluorescence is a means for visualizing and measuring the local concentration of such species. A number of respective experiments have been performed in which use has been made of the fluorescent properties of the radical OH (Dyer and Crosley, 1982; Kychakoff *et al.*, 1983, 1984a; Cattolica and Vosen, 1986), and nitric oxide can be used as well for this kind of experiments (Kychakoff *et al.*, 1984b). The resonant wavelengths of OH used for excitation are in the visible and near ultraviolet part of the spectrum. Since it is of greatest interest to perform measurements with turbulent flames, a pulsed dye laser should be used as the exciting light source. The laser beam, whose wavelength is tuned to a resonance wavelength, is expanded to illuminate a thin, plane sheet in the flame. The incident radiation excites a higher electronic energy level of the radical OH. The intensity of the emitted fluorescent signal, expressed, e.g., as the number of photons striking the receiver, is proportional to the number of density of OH in the scattering volume (i.e., in the laser sheet). The factor of this proportionality might depend on the temperature and on certain system parameters. The signal intensity is much lower than in the

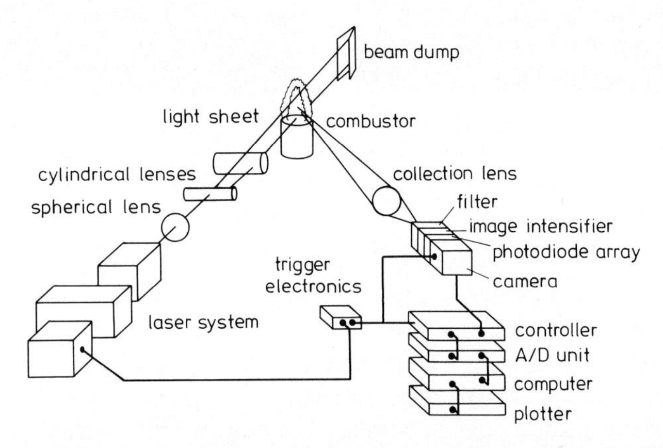

Fig. 4.12 Schematic arrangement for laser-induced fluorescence visualization in a flame. The light pulse from a dye laser is expanded to illuminate a thin, plane sheet in the flame. The OH fluorescence is intensified and recorded with a photodiode array that is connected to a data evaluation system. (From Kychakoff *et al.*, 1984a. Copyright, Optical Society of America.)

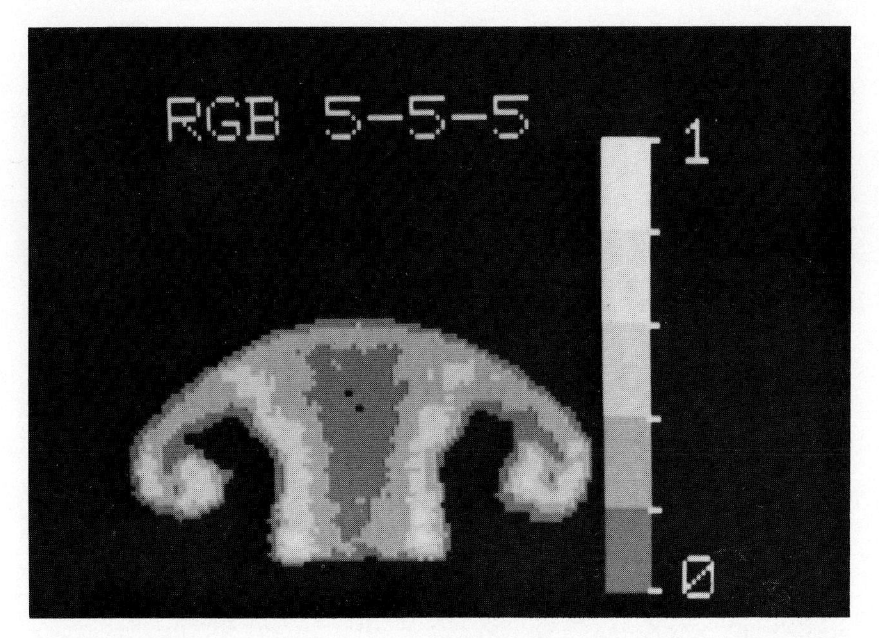

Fig. 4.13 Digital image representation of the measured OH concentration in a flame-vortex interaction. Original figure is in color and is the result of a planar laser-induced fluorescence measurement. (From Cattolica and Vosen, 1986.)

fluorescence experiments performed with sodium or iodine tracers, particularly when the signal is generated with a short laser pulse. It is therefore necessary to apply an image intensifier whose output can be given on an array of photodetectors or a vidicon camera (Fig. 4.12). The recorded pattern is digitized and can be displayed, thus allowing for a visualization of small and large scale structures in the flame (Fig. 4.13). Different values of the relative OH-concentration can be attributed to the individual gray levels of such a picture. The measurement of the absolute concentration would require to perform at least one single-point calibration by means of another technique.

References

Ackermann, U., Baganoff, D., and McDaniel, J. C. (1985). Dependence of laser-induced fluorescence on gas-dynamic fluctuations with application to measurements in unsteady flows. *Exp. Fluids* **3,** 45–51.

Asanuma, T., Tanida, Y., and Kurihara, K. (1979). On the measurement of flow velocity by means of spark tracing method. *In* "Flow Visualization" (T. Asanuma, ed.), pp. 227–232. Hemisphere, Washington, D.C.

Bernotat, S., and Umhauer, H. (1973). Application of "spark tracing" method to flow measurements in an air classifier. *Opto-electronics (London)* **5,** 107–118.

Bomelburg, H. J., Herzog, J. R., and Weske, J. R. (1959). The electric spark method for quantitative measurements in flowing gases. *Z. Flugwiss.* **7,** 322–329.

Brydon, J. W. E., Lambie, A. K., and Wheatley, D. J. (1979). Thermographicvisualization of coronary artery blood flow during by-pass surgery. *J. Med. Eng. Technol.* **3,** 77–80.

Cattolica, R. J., and Vosen, S. R. (1986). Fluorescence imaging of a flame-vortex interaction. *Comb. Sci. Technol.* **48,** 77–88.

Cenkner, A. A., and Driscoll, R. J. (1982). Laser-induced fluorescence visualization on supersonic mixing nozzles that employ gas-trips. *AIAA J.* **20,** 812–819.

Dewey, J. M. (1973). The analysis of the particle trajectories in unsteady shock flows. *Proc. Int. Congr. Instrum. Aerosp. Simul. Facilities, 5th,* pp. 119–124.

Dickerson, R. R., and Stedman, D. H. (1982). Ozone flow visualization techniques. *In* "Flow Visualization II" (W. Merzkirch, ed.), pp. 677–681. Hemisphere, Washington, D.C.

Dyer, M. J., Crosley, D. R. (1982). Two-dimensional imaging of OH laser-induced fluorescence in a flame. *Opt. Lett.* **7,** 382–384.

Fister, W. (1966). Sichtbarmachung der Strömungen in Radialverdichterstufen, besonders der Relativströmung in rotierenden Laufrädern, durch Funkenblitze. *Brennst.-Waerme-Kraft* **18,** 425–429.

Fister, W., Eickelmann, J., and Witzel, U. (1982). Expanded application programs of the spark tracer method with regard to centrifugal compressor impellers. *In* "Flow Visualization II" (W. Merzkirch, ed.), pp. 107–120. Hemisphere, Washington, D.C.

Früngel, F. (1960). Bewegungsaufnahmen rascher Luftströmungen und Stosswellen durch hochfrequente Hochspannungsfunken. *Jahrb. Wiss. Ges. Luft-Raumfahrt.* **1960,** 175–182.

Früngel, F. (1977). High-frequency spark tracing and application in engineering and aerodynamics. *Proc. Int. Congr. High-Speed Photogr., 12th*; *Proc. Soc. Photo-Opt. Instrum. Eng.* **97,** 291–301.

Früngel, F., and Thorwart, W. (1970). Spark tracing method progress in the analysis of gaseous flows. *Proc. Int. Congr. High-Speed Photogr., 9th,* pp. 166–170.

Grün, A. E., Schopper, E., and Schumacher, B. (1953). Electron shadowgraph and afterglow pictures of gas jets at low densities. *J. Appl. Phys.* **24,** 1527–1528.

Hartunian, R. A., and Spencer, D. J. (1966). Visualization technique for massive blowing studies. *AIAA J.* **4,** 1305–1307.

Hiller, B., and Hanson, R. K. (1985). Two-frequency laser-induced fluorescence technique for rapid velocity field measurements in gas flows. *Opt. Lett.* **10,** 206–209.

Hiller, W. J., and Hägele, J. (1982). Visualization of hypersonic micro-jets by laser-induced fluorescence. *In* "Flow Visualization II" (W. Merzkirch, ed.), pp. 427–431. Hemisphere, Washington, D.C.

Horstman, C. C., and Kussoy, M. I. (1968). Hypersonic viscous interaction on slender cones. *AIAA J.* **6,** 2364–2371.

Kimura, T., Nishio, M., Fujita, T., and Maeno, R. (1977). Visualization of shock wave by electric discharge. *AIAA J.* **15,** 611–612.

Koochesfahani, M. M., and Dimotakis, P. E. (1985). Laser-induced fluorescence measurements of mixed fluid concentration in a liquid plane shear layer. *AIAA J.* **23,** 1700–1707.

Kychakoff, G., Howe, R. D., Hanson, R. K., and Knapp, K. (1983). Flow visualization in combustion gases. *AIAA Pap.* **83-0405.**

Kychakoff, G., Howe, R. D., and Hanson, R. K. (1984a). Quantitative flow visualization technique for measurements in combustion gases. *Appl. Opt.* **23,** 704–712.

Kychakoff, G., Knapp, K., Howe, R. D., and Hanson, R. K. (1984b). Flow visualization in combustion gases using nitric oxide fluorescence. *AIAA J.* **22,** 153–154.

Kyser, J. B. (1964). Tracer-spark technique for velocity mapping of hypersonic flow fields. *AIAA J.* **2,** 393–394.

Lahaye, C., Leger, E. G., and Lemay, A. (1967). Wake velocity measurements using a sequence of sparks. *AIAA J.* **5,** 2274–2276.

Léwy, S. (1970). Visualisations d'écoulements en soufflerie à l'aide d'un faisceau d'électrons. *Rech. Aerosp.* No. **1970-3,** 155–166.

Matsuo, K., Ikui, T., Yamamoto, Y., and Setoguchi, T. (1979). Measurements of shock tube flows using a spark tracer method. *In* "Flow Visualization" (T. Asanuma, ed.), pp. 233–238. Hemisphere, Washington, D.C.

Matsuo, K., Setoguchi, T., and Yamamoto, Y. (1981). The error in measuring an accelerated flow velocity by a spark tracer method. *Bull. JSME* **24,** 1168–1175.

McCroskey, W. J., Bogdonoff, S. M., and McDougall, J. G. (1966). An experimental model for the sharp flat plate in rarefied hypersonic flow. *AIAA J.* **4,** 1580–1587.

McDaniel, J. C. (1983). Quantitative measurement of density and velocity in compressible flows using laser-induced fluorescence. *AIAA Pap.* **83-0049.**

McDaniel, J. C., Baganoff, D., and Byer, R. L. (1982). Density measurements in compressible flows using off-resonant laser-induced fluorescence. *Phys. Fluids* **25,** 1105–1107.

McIntoch, M. K. (1971). Free stream velocity measurements in a high enthalpy shock tunnel. *Phys. Fluids* **14,** 1100–1102.

Miles, R. B., Udd, E., and Zimmerman, M. (1978). Quantitative flow visualization in sodium vapor seeded hypersonic helium. *Appl. Phys. Lett.* **32,** 317–319.

Muntz, E. P. (1968). The electron beam fluorescence technique. *AGARDograph* **AGARD-AG-132.**

Muntz, E. P. (1981). Measurement of density by analysis of electron beam excited radiation. *In* "Fluid Dynamics" (R. J. Emrich, ed.), Methods of Experimental Physics, Vol. 18. pp. 434–455. Academic Press, New York.

Nakayama, Y., Okitsu, S., Aoki, K., and Ohta, H. (1979). Flow direction detectable spark method. *In* "Flow Visualization" (T. Asanuma, ed.), pp. 239–244. Hemisphere, Washington, D.C.

Ohashi, H., and Ishikawa, N. (1972). Visualization study of flow near the trailing edge of an oscillating airfoil. *Bull. JSME* **15,** 840–847.

Pierce, D. (1961). Photographic evidence of the formation and growth of vorticity behind plates accelerated from rest in still air. *J. Fluid Mech.* **11,** 460–464.

Rapagnani, N. L., and Davis, S. J. (1979). Laser-induced I_2 fluorescence measurements in a chemical laser flowfield. *AIAA J.* **17,** 1402–1404.

Rothe, D. E. (1965). Flow visualization using a traversing electron beam. *AIAA J.* **3,** 1945–1946.

Rudinger, G., and Somers, L. M. (1960). Behaviour of small regions of different gases carried in accelerated gas flows. *J. Fluid Mech.* **7,** 161–176.

San, W., and Ge, Y. (1982). Two methods for low density flow visualization. *In* "Flow Visualization II" (W. Merzkirch, ed.), pp. 421–425. Hemisphere, Washington, D.C.

Schardin, H. (1942). Die Schlierenverfahren und ihre Anwendungen. *Ergeb. Exakten Naturwiss.* **20,** 303–439.

Schumacher, B. (1953). Abbildung von Gasströmungen mit Elektronenstrahlen. *Ann. Phys.* **6,** 404–420.

Schumacher, B., and Gadamer, E. O. (1958). Electron beam fluorescence probe for measuring the local gas density in a wide field of observation. *Can. J. Phys.* **36,** 659–671.

Sebacher, D. I. (1966). Flow visualization using an electron-beam afterglow in N_2 and air. *AIAA J.* **4,** 1858–1859.

Van der Bliek, J. A., Cassanova, R. A., Golomb, D., Del Greco, F. P., Hill, J. A. F., and Good, R. E. (1967). The chemiluminescent reaction of NO with O in a supersonic low density wind tunnel. *In* "Rarefied Gas Dynamics" (C. L. Brundin, ed.), pp. 1543–1560. Academic Press, New York.

Weinstein, L. M., Wagner, R. D., and Ocheltree, S. L. (1968). Electron beam flow visualization in hypersonic helium flow. *AIAA J.* **6,** 1623–1625.

Winkler, E. M. (1954). Electrical discharge and afterglow technique. *In* "Physical Measurements in Gas Dynamics and Combustion" (R. W. Ladenburg, ed.), pp. 79–88. Princeton Univ. Press, Princeton, New Jersey.

Zimmermann, M., Cheng, S., and Miles, R. B. (1985). Velocity selective flow visualization in a free supersonic nitrogen jet with the resonant Doppler velocimeter. *In* "Flow Visualization III" (W.-J. Yang, ed.), pp. 449–453. Hemisphere, Washington, D.C.

Index

E

F

G

H

I

K

L

M

T

V

W

Y